The Anatomy
Student's Self-test
Coloring Book

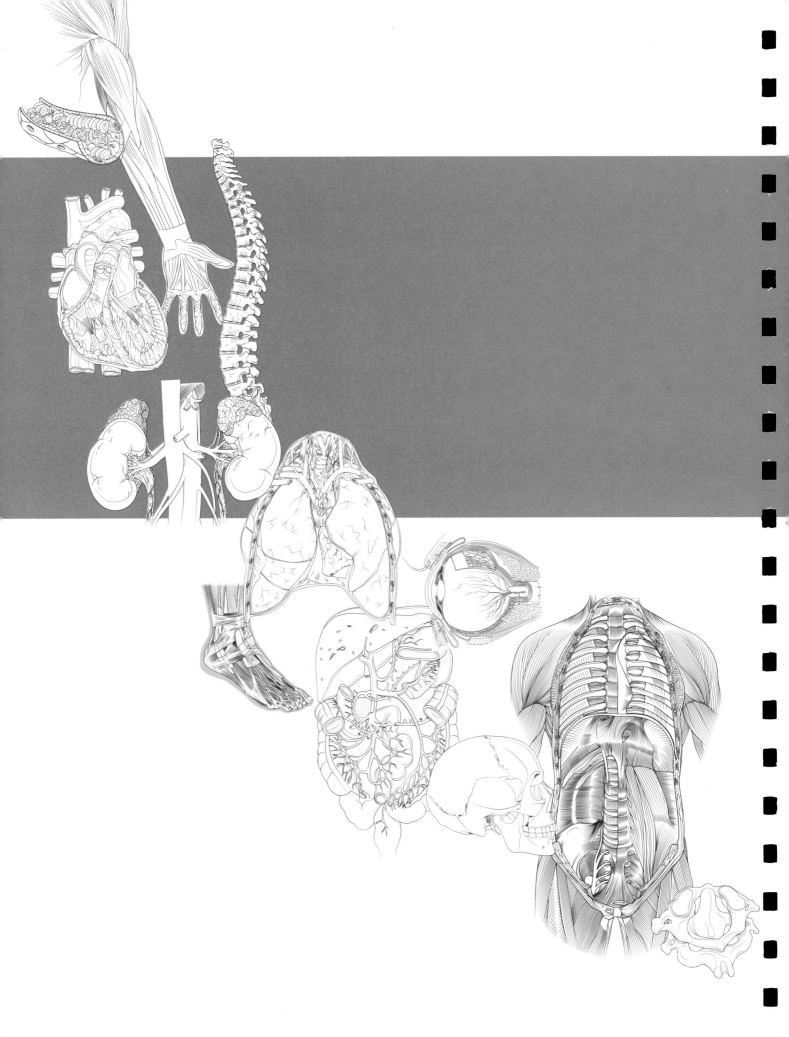

The Anatomy
Student's Self-test
Coloring Book

Chief Consultant: **Kurt H. Albertine**, Ph.D.

University of Utah, School of Medicine

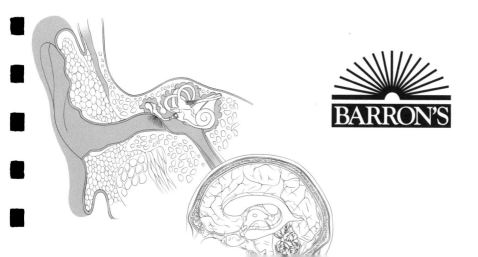

BARRON'S

First edition for the United States, its territories, and Canada
published in 2007 by Barron's Educational Series, Inc.

Reprinted in 2009

First published in 2007 by Global Book Publishing Pty Ltd
This publication and arrangement © Global Book Publishing Pty Ltd 2007
Text © Global Book Publishing Pty Ltd 2007
Illustrations © Global Book Publishing Pty Ltd 2007

All inquiries should be addressed to:
Barron's Educational Series, Inc.
250 Wireless Boulevard
Hauppauge, New York 11788
www.barronseduc.com

ISBN-13: 978-0-7641-3777-8
ISBN-10: 0-7641-3777-8

MANAGING DIRECTOR	Chryl Campbell
PUBLISHING DIRECTOR	Sarah Anderson
ART DIRECTOR	Kylie Mulquin
PROJECT MANAGER	Kate Etherington
CHIEF CONSULTANT	Kurt H. Albertine, PhD
CONTRIBUTORS	Robin Arnold, MSc
	Ken Ashwell, BMedSc, MB, BS, PhD
	Deborah Bryce, BSc, MScQual, MChiro, GrCertHEd
	John Gallo, MB, BS(Hons), FRACP, FRCPA
	Rakesh Kumar, MB, BS, PhD
	Peter Lavelle, MB, BS
	Karen McGhee, BSc
	Michael Roberts, MB, BS, LLB(Hons)
	Emeritus Professor Frederick Rost, BSc(Med), MB, BS, PhD, DCP(London), DipRMS
	Elizabeth Tancred, BSc, PhD
	Dzung Vu, MD, MB, BS, DipAnat, GradCertHEd
	Phil Waite, BSc(Hons), MBChB, CertHEd, PhD
EDITORIAL CONSULTANT	Janet Parker
EDITOR	Kate Etherington
PROOFREADER	Elizabeth Connolly
ILLUSTRATION EDITORS	Alan Edwards
	Heather McNamara
ILLUSTRATORS	Mike Gorman
	Thomson Digital
	Glen Vause
COVER DESIGN	Stan Lamond
	Kylie Mulquin
DESIGN CONCEPT AND LAYOUT	Stan Lamond
INDEX	Jon Jermey
PRODUCTION MANAGER	Ian Coles
RIGHTS MANAGER	Belinda Vance
PUBLISHING ASSISTANT	Jessica Luca

Printed in China by SNP Leefung Printers Limited
Color separation Pica Digital Pte Ltd, Singapore
9 8 7 6 5 4 3 2

Contents

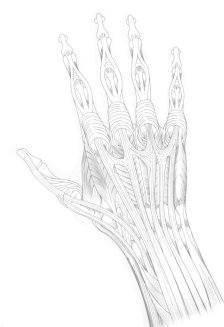

Introduction

The structures and systems of the human body have fascinated us for millennia. As early as the seventeenth century BCE, the Ancient Egyptians were making detailed notes about the parts of the body and their relation to each other. Leonardo da Vinci's anatomical drawings in the late 1400s and early 1500s certainly advanced the study of anatomy, despite the horror with which dissection was viewed at the time. Today, many museums and popular traveling exhibitions display real human specimens, allowing the general public to gain fascinating insight into how the human body is constructed.

Many resources are available to medical students, doctors, nurses, health professionals, fitness trainers, nutritionists, artists, and others who desire to learn about human anatomy, or who need to brush up on their knowledge of the human body from time to time. What sets this book apart from the exhibitions, Internet sites, CD-ROMs, and other material is that it is an *active* learning experience—by coloring in the parts of the body, the shape and location of each part becomes firmly fixed in your mind, and writing the name of each part in the space provided and checking it against the answers at the bottom of the page impresses the word and its spelling on the mind far more readily than if you had simply viewed a diagram in a textbook. The unique connection between hand, eye, and mind makes this anatomy coloring book an invaluable study tool for people of all ages and education levels.

How This Book is Organized

Featuring more than 350 computer-rendered line drawings in a clean design, this book is divided into 16 comprehensive chapters. It covers the major body systems— from the skeletal and muscular to the lymphatic and circulatory—and also includes an overview of the body, cells and tissues, articulations, special sense organs, and human development. Each of the eight full-color acetate sheets inserted at the front of the book shows a whole body system or part of one; cut these out and place them over the line drawings of body systems within the book to reveal how different systems interact. This is an important facet in the study of anatomy—not only learning the names and locations of the parts of the body, but also discovering how these parts relate to each other and function together.

As you work your way through this book, you will gain both a clear understanding of anatomy and a deeper appreciation for the human body—an amazingly complicated yet perfectly coordinated organic machine.

Kurt H. Albertine

How to Use This Book

This book is designed to assist students and professionals to identify body parts and structures, and the colored leader lines aid the process by clearly pointing out each body part. The functions of coloring and labeling allow you to familiarize yourself with individual parts of the body and then check your knowledge.

Coloring is best done using either pencils or ballpoint pens (not felt-tip pens) in a variety of dark and light colors. Where possible, you should use the same color for like structures, so that all completed illustrations can be utilized later as visual references. According to anatomical convention, the color green is usually reserved for lymphatic structures, yellow for nerves, red for arteries, and blue for veins.

Labeling the colored leader lines that point to separate parts of the illustration enables you to test and then check your knowledge using the answers that are printed at the bottom of the page.

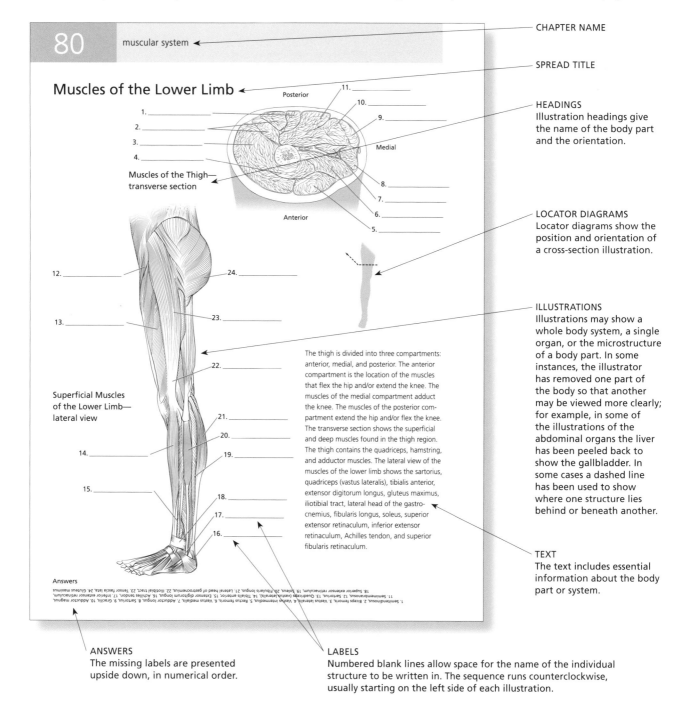

CHAPTER NAME

SPREAD TITLE

HEADINGS
Illustration headings give the name of the body part and the orientation.

LOCATOR DIAGRAMS
Locator diagrams show the position and orientation of a cross-section illustration.

ILLUSTRATIONS
Illustrations may show a whole body system, a single organ, or the microstructure of a body part. In some instances, the illustrator has removed one part of the body so that another may be viewed more clearly; for example, in some of the illustrations of the abdominal organs the liver has been peeled back to show the gallbladder. In some cases a dashed line has been used to show where one structure lies behind or beneath another.

TEXT
The text includes essential information about the body part or system.

ANSWERS
The missing labels are presented upside down, in numerical order.

LABELS
Numbered blank lines allow space for the name of the individual structure to be written in. The sequence runs counterclockwise, usually starting on the left side of each illustration.

Overview of the Human Body

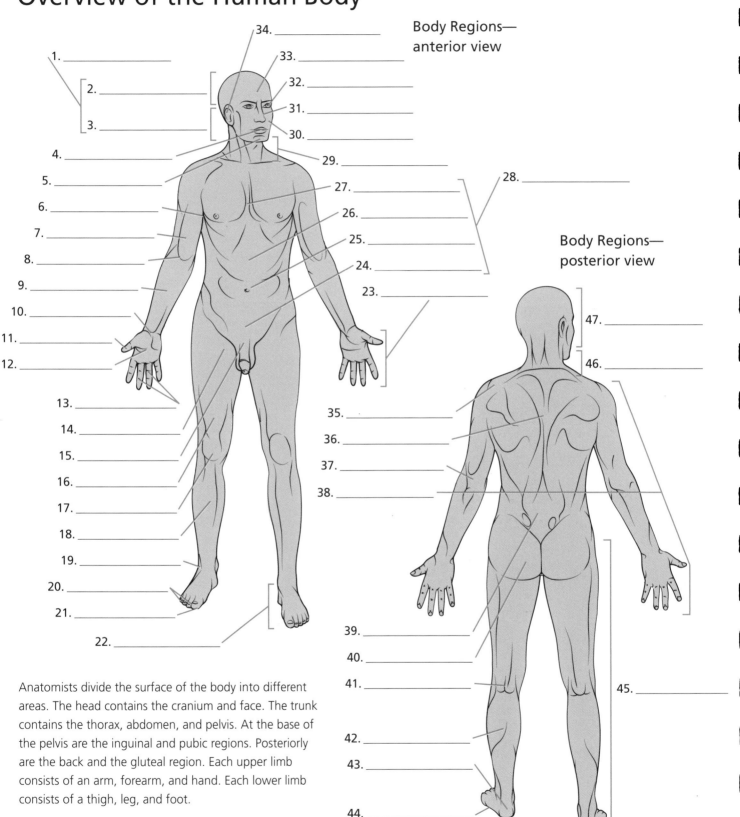

Body Regions—
anterior view

1. _____
2. _____
3. _____
4. _____
5. _____
6. _____
7. _____
8. _____
9. _____
10. _____
11. _____
12. _____
13. _____
14. _____
15. _____
16. _____
17. _____
18. _____
19. _____
20. _____
21. _____
22. _____

34. _____
33. _____
32. _____
31. _____
30. _____
29. _____
27. _____
26. _____
25. _____
24. _____
23. _____

28. _____

Body Regions—
posterior view

47. _____
46. _____

35. _____
36. _____
37. _____
38. _____

39. _____
40. _____
41. _____
42. _____
43. _____
44. _____

45. _____

Anatomists divide the surface of the body into different areas. The head contains the cranium and face. The trunk contains the thorax, abdomen, and pelvis. At the base of the pelvis are the inguinal and pubic regions. Posteriorly are the back and the gluteal region. Each upper limb consists of an arm, forearm, and hand. Each lower limb consists of a thigh, leg, and foot.

Answers

1. Head, 2. Cranium (cranial), 3. Face (facial), 4. Mouth (oral), 5. Chin (mental), 6. Axilla (axillary), 7. Brachium (brachial), 8. Elbow (antecubital), 9. Antebrachium (antebrachial), 10. Wrist (carpal), 11. Pollex (thumb), 12. Palm (palmar), 13. Digits (digital or phalangeal), 14. Inguen (inguinal), 15. Pubis (pubic), 16. Femur (femoral), 17. Patella (patellar), 18. Crus (crural), 19. Tarsus (tarsal), 20. Digits (digital or phalangeal), 21. Hallux (big toe), 22. Pes (pedal), 23. Hand, 24. Pelvis (pelvic), 25. Umbilicus (umbilical), 26. Abdomen (abdominal), 27. Thorax (thoracic), 28. Trunk, 29. Neck (cervical), 30. Cheek (buccal), 31. Nose (nasal), 32. Eye (orbital or ocular), 33. Forehead (frontal), 34. Ear (otic), 35. Shoulder (acromial), 36. Back (dorsal), 37. Olecranon (olecranal), 38. Upper limb, 39. Lower back (lumbar), 40. Gluteus (gluteal), 41. Popliteus (popliteal), 42. Sura (sural), 43. Calcaneus (calcaneal), 44. Sole (plantar), 45. Lower limb, 46. Neck (cervical), 47. Head

The body cavities are spaces containing the internal organs (viscera). The main cavities are the thoracic and abdominopelvic cavities in the torso and the cranial cavity in the head. The thoracic (or chest) cavity contains the heart, lungs, trachea, and esophagus. The abdominopelvic cavity is divided into the abdominal cavity, which contains most of the gastrointestinal tract, the kidneys, and the adrenal glands; below the abdominal cavity, the pelvic cavity contains the urogenital system and the rectum. The cranial cavity contains the brain and extends caudally as the spinal canal.

Body Cavities—
sagittal view

10. *dorsal cavity*

1. *torso*

2. *thoracic cavity*

3. *pericardial cavity*

5. *diaphragm*

4. *Abdominopelvic cavity*

6. *abdominal cavity*

7. *pelvic cavity*

9. *cranial cavity*

8. *spinal canal*

View Orientation and Anatomical Planes

Specific terms describe the orientation and relationships of the body and its parts. Sections of the body are described in terms of anatomical planes (flat surfaces). These are imaginary lines—vertical or horizontal—drawn through a body in the anatomical position (that is, with the body standing erect, feet together with toes pointed forward, and arms at the sides with palms facing forward). A transverse (axial or horizontal) plane cuts the body across from side to side, separating superior areas above from inferior areas below. A coronal (frontal) plane divides the body into dorsal (posterior, or back) and ventral (anterior, or front) pieces. A sagittal plane separates one side of the body from the other side (left from right). The midsagittal (median) plane is the sagittal plane exactly in the middle of the body. The relationships of one body part to another are identified by terms such as medial (toward the midline of the body) or lateral (away from the midline of the body); inferior (below, or lower) or superior (upper, or above); cranial (toward the head) or caudal (toward the tail); anterior (ventral, or toward the front) or posterior (dorsal, or toward the back); proximal (closer to a reference point) or distal (farther from that reference point).

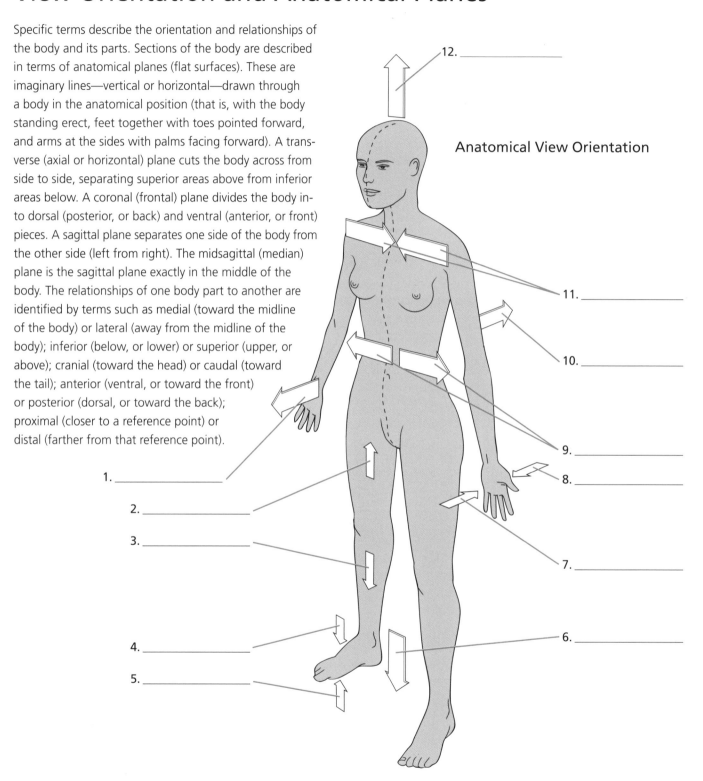

Anatomical View Orientation

12. _____

11. _____

10. _____

9. _____

8. _____

7. _____

6. _____

1. _____

2. _____

3. _____

4. _____

5. _____

Answers

1. Anterior, 2. Proximal, 3. Distal, 4. Dorsal, 5. Plantar, 6. Inferior, 7. Palmar, 8. Dorsal, 9. Lateral, 10. Posterior, 11. Medial, 12. Superior

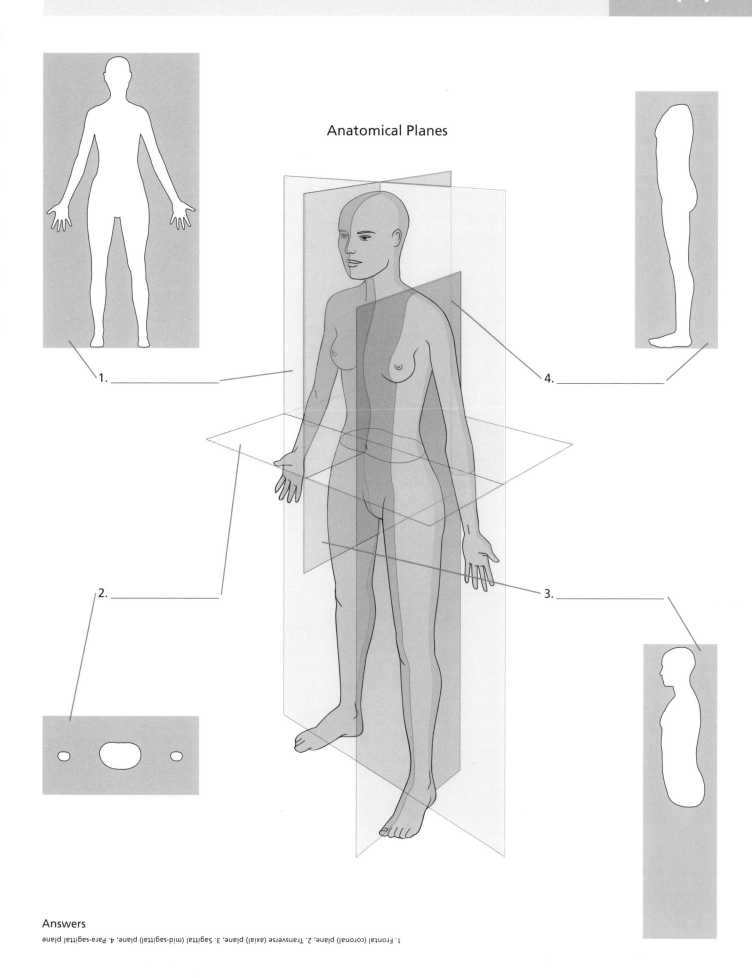

Anatomical Planes

1. _____

2. _____

3. _____

4. _____

Answers

1. Frontal (coronal) plane, 2. Transverse (axial) plane, 3. Sagittal (mid-sagittal) plane, 4. Para-sagittal plane

Cell Structure and Major Cell Types

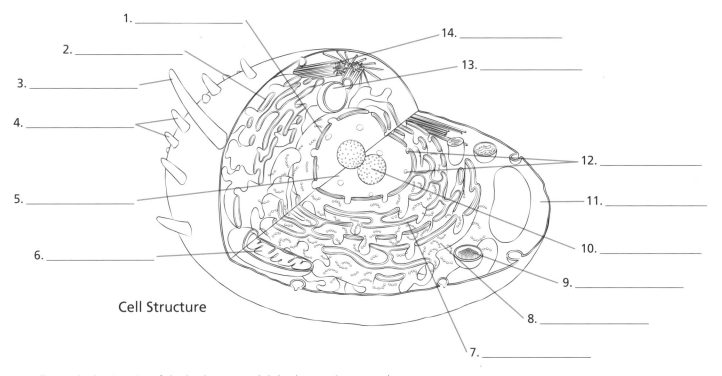

1. _____
2. _____
3. _____
4. _____
5. _____
6. _____

14. _____
13. _____
12. _____
11. _____
10. _____
9. _____
8. _____
7. _____

Cell Structure

Cells are the basic units of the body. Every adult body contains more than 5 trillion cells. Cells are surrounded by a cell membrane. Within the cell membrane lies the cytoplasm, a fluid containing many important structural units called organelles. These include rough endoplasmic reticulum, mitochondria, Golgi apparatus, and centrioles. The nucleus is separate from the cytoplasm. Cells are specialized to perform particular functions. Neurons are specialized cells that conduct nerve impulses. Each has three main parts: the cell body, branching projections (dendrites) that carry impulses to the cell body, and one elongated projection (axon) that conveys impulses away from the cell body.

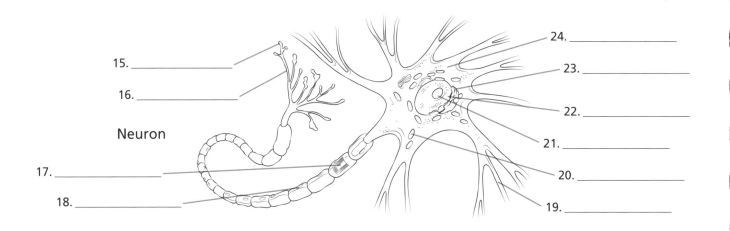

15. _____
16. _____

Neuron

17. _____
18. _____

24. _____
23. _____
22. _____
21. _____
20. _____
19. _____

Answers

1. Nucleus, 2. Golgi apparatus, 3. Cilium, 4. Microvilli, 5. Location of chromatin, 6. Mitochondrion, 7. Endoplasmic reticulum, 8. Ribosome, 9. Peroxisome, 10. Nucleolus, 11. Cytoplasm, 12. Nuclear pores, 13. Lysosome, 14. Centriole, 15. Synaptic knob, 16. Axon terminal, 17. Axon, 18. Myelin sheath, 19. Dendrite, 20. Mitochondrion, 21. Nucleolus, 22. Nuclear membrane, 23. Golgi apparatus, 24. Cell body

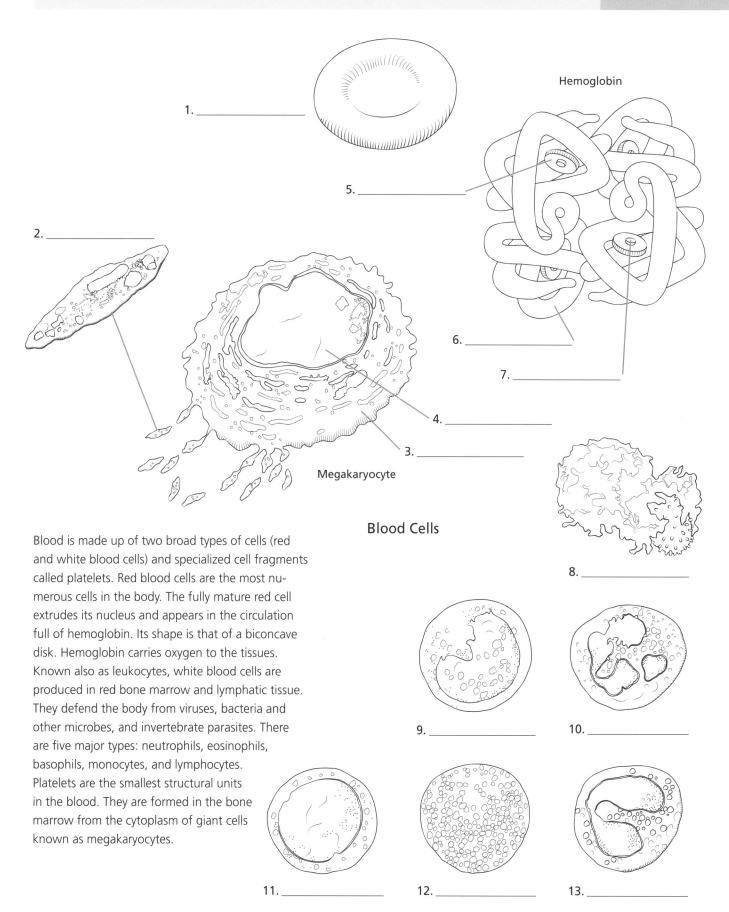

Hemoglobin

1. _____

2. _____

5. _____

6. _____

7. _____

4. _____

3. _____

Megakaryocyte

Blood Cells

Blood is made up of two broad types of cells (red and white blood cells) and specialized cell fragments called platelets. Red blood cells are the most numerous cells in the body. The fully mature red cell extrudes its nucleus and appears in the circulation full of hemoglobin. Its shape is that of a biconcave disk. Hemoglobin carries oxygen to the tissues. Known also as leukocytes, white blood cells are produced in red bone marrow and lymphatic tissue. They defend the body from viruses, bacteria and other microbes, and invertebrate parasites. There are five major types: neutrophils, eosinophils, basophils, monocytes, and lymphocytes. Platelets are the smallest structural units in the blood. They are formed in the bone marrow from the cytoplasm of giant cells known as megakaryocytes.

8. _____

9. _____

10. _____

11. _____

12. _____

13. _____

Answers

Tissues (cellular level)

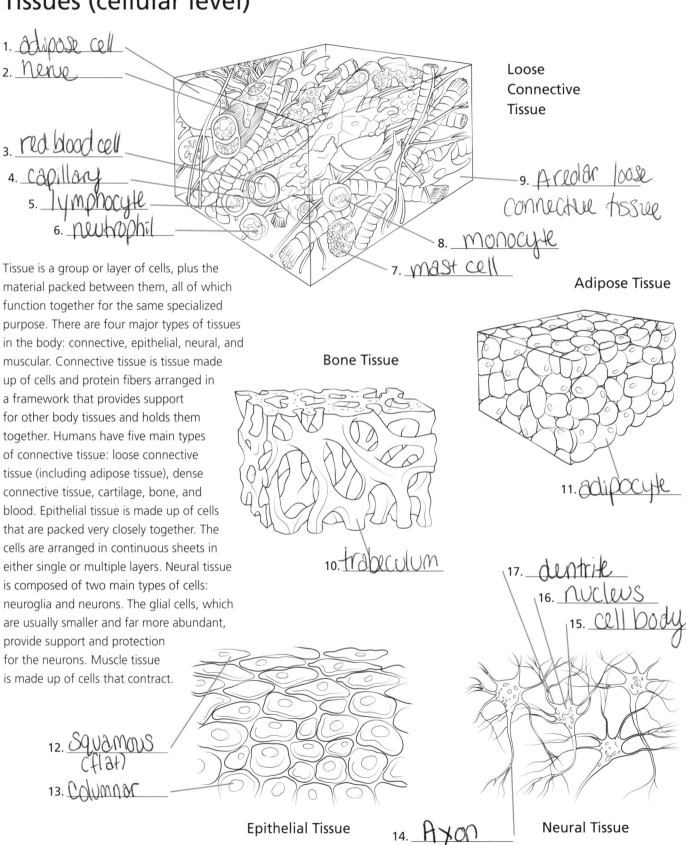

1. adipose cell
2. Nerve

Loose Connective Tissue

3. red blood cell
4. capillary
5. lymphocyte
6. neutrophil

9. Areolar loose connective tissue

8. monocyte

7. mast cell

Tissue is a group or layer of cells, plus the material packed between them, all of which function together for the same specialized purpose. There are four major types of tissues in the body: connective, epithelial, neural, and muscular. Connective tissue is tissue made up of cells and protein fibers arranged in a framework that provides support for other body tissues and holds them together. Humans have five main types of connective tissue: loose connective tissue (including adipose tissue), dense connective tissue, cartilage, bone, and blood. Epithelial tissue is made up of cells that are packed very closely together. The cells are arranged in continuous sheets in either single or multiple layers. Neural tissue is composed of two main types of cells: neuroglia and neurons. The glial cells, which are usually smaller and far more abundant, provide support and protection for the neurons. Muscle tissue is made up of cells that contract.

Adipose Tissue

11. adipocyte

Bone Tissue

10. trabeculum

17. dentrite
16. nucleus
15. cell body

12. Squamous (flat)
13. Columnar

Epithelial Tissue

14. Axon

Neural Tissue

Muscle Tissue

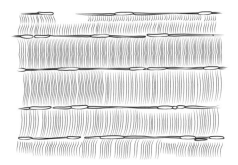

1. Skeletal muscle

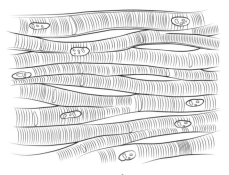

2. Cardiac muscle

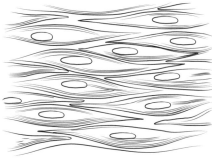

3. Smooth muscle

Cartilage Tissue

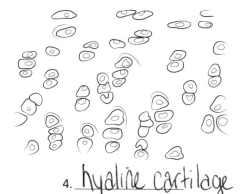

4. hyaline cartilage

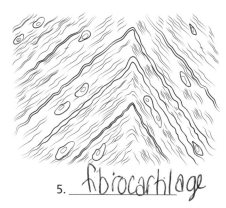

5. fibrocartilage

6. elastic cartilage

Dense Regular Connective Tissue

Muscle tissue is composed of cells that are purpose-built for contraction. There are three main forms: skeletal, cardiac, and smooth. The muscle fibers, or myofibers, of skeletal muscle are long and cylindrical, are arranged parallel to each other, and have a striped appearance under the microscope. Cardiac muscle cells are similar in that they too have a striped appearance. They, however, are branched. Smooth muscle tissue has no striations. One type of dense connective tissue, known as dense regular connective tissue, forms tendons, ligaments, and cartilage. Another, elastic connective tissue, is specialized for stretching. Cartilage comes in three forms: hyaline cartilage, fibrocartilage, and elastic cartilage. Hyaline cartilage covers the bones where they form synovial joints. Fibrocartilage forms a component of some other joints, which usually have a limited range of movement. Elastic cartilage is the hard material that can be felt in the external ear.

7. tendon tissue

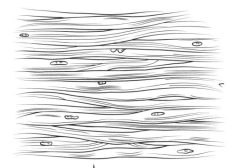

8. ligament tissue

Answers

Integumentary System

The skin is a protective organ that covers the body, merging with mucous membranes at the openings of the body such as the mouth and anus. Specialized nerve receptors in the skin also allow the body to sense pain, hot and cold, touch, and pressure. Skin is composed of two layers: the epidermis and the dermis. The epidermis is the outer protective layer. The deepest layers of the epidermis are the basal and spinous layers (the stratum granulosum and stratum spinosum, respectively); the outer layer is the horny layer (or stratum corneum). The dermis is the skin's inner layer and contains networks of blood vessels and nerves. Beneath the dermis is a layer of fat cells. Among the layers of the epidermis and the dermis are the specialized structures—hair follicles, sweat glands, and sebaceous glands. Hair and nails are made mainly from a tough hard protein called keratin produced in the deeper layers of the skin's epidermis.

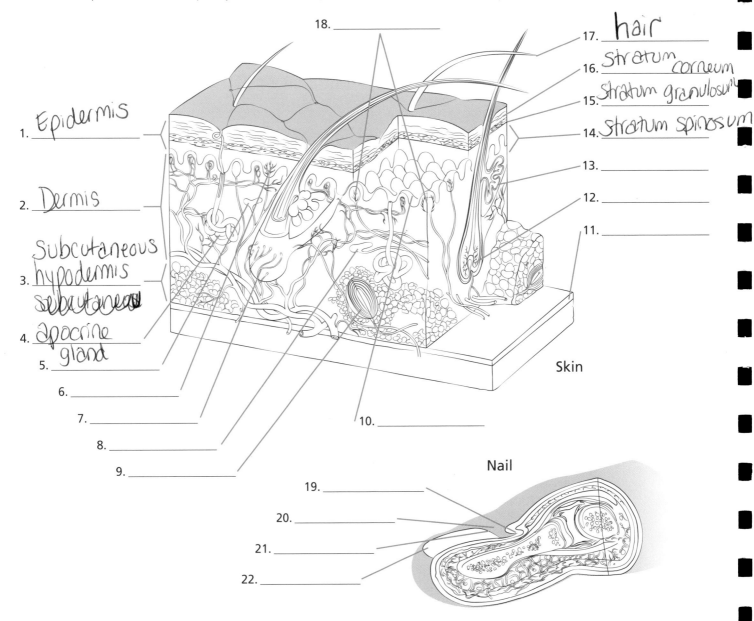

18. _____

17. _hair_

16. _Stratum corneum_

15. _Stratum granulosum_

14. _Stratum spinosum_

1. _Epidermis_

13. _____

12. _____

11. _____

2. _Dermis_

3. _Subcutaneous hypodermis_

sebataneous

4. _apocrine gland_

5. _____

6. _____

7. _____

8. _____

9. _____

10. _____

Skin

Nail

19. _____

20. _____

21. _____

22. _____

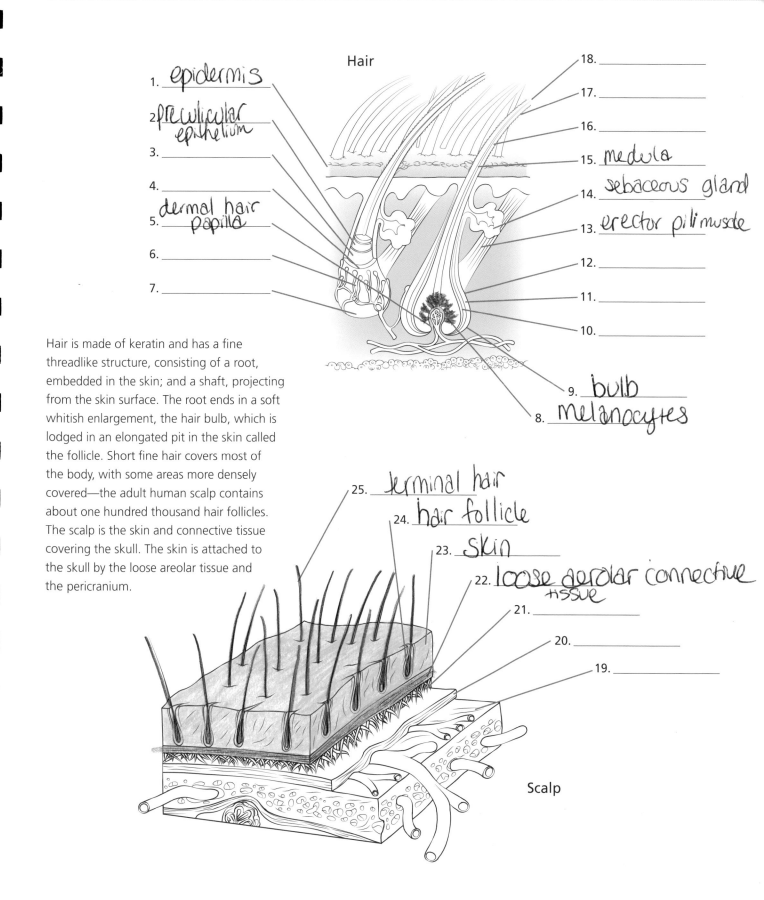

Hair

1. epidermis
2. precuticular epithelium
3. _____
4. _____
5. dermal hair papilla
6. _____
7. _____

18. _____
17. _____
16. _____
15. medula
14. sebaceous gland
13. erector pili muscle
12. _____
11. _____
10. _____

9. bulb
8. Melanocytes

Hair is made of keratin and has a fine threadlike structure, consisting of a root, embedded in the skin; and a shaft, projecting from the skin surface. The root ends in a soft whitish enlargement, the hair bulb, which is lodged in an elongated pit in the skin called the follicle. Short fine hair covers most of the body, with some areas more densely covered—the adult human scalp contains about one hundred thousand hair follicles. The scalp is the skin and connective tissue covering the skull. The skin is attached to the skull by the loose areolar tissue and the pericranium.

25. terminal hair
24. hair follicle
23. skin
22. loose aerolar connective tissue
21. _____
20. _____
19. _____

Scalp

Skeletal System

The clavicle + scapula are collectively known as the pectoral girdles.

Skeletal System— anterior view

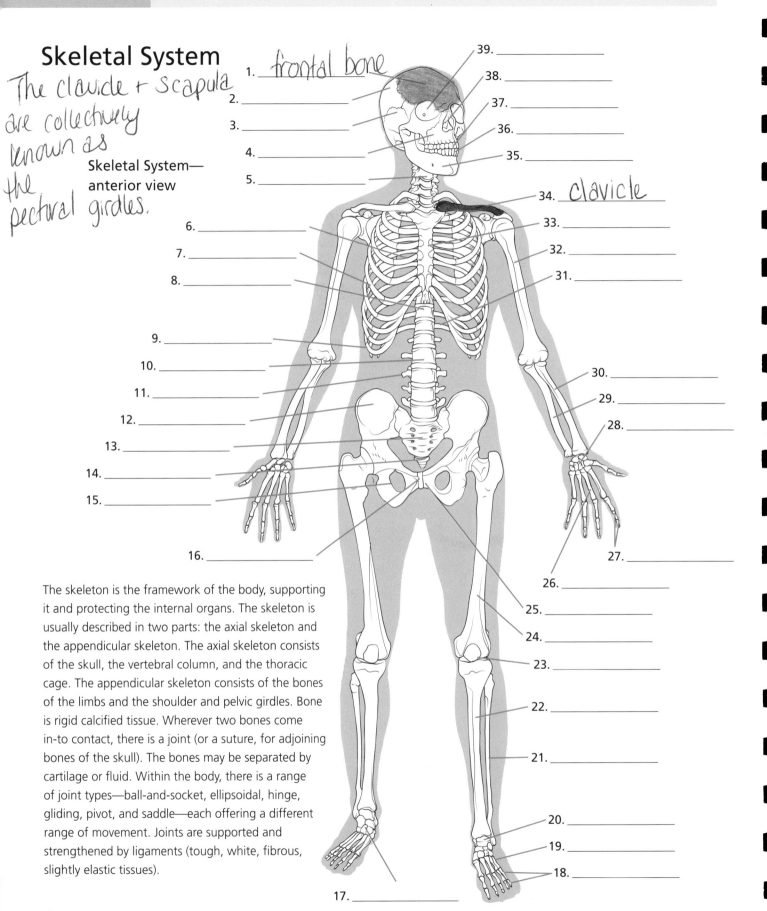

1. frontal bone
2. _____
3. _____
4. _____
5. _____
6. _____
7. _____
8. _____
9. _____
10. _____
11. _____
12. _____
13. _____
14. _____
15. _____
16. _____
17. _____
18. _____
19. _____
20. _____
21. _____
22. _____
23. _____
24. _____
25. _____
26. _____
27. _____
28. _____
29. _____
30. _____
31. _____
32. _____
33. _____
34. clavicle
35. _____
36. _____
37. _____
38. _____
39. _____

The skeleton is the framework of the body, supporting it and protecting the internal organs. The skeleton is usually described in two parts: the axial skeleton and the appendicular skeleton. The axial skeleton consists of the skull, the vertebral column, and the thoracic cage. The appendicular skeleton consists of the bones of the limbs and the shoulder and pelvic girdles. Bone is rigid calcified tissue. Wherever two bones come in-to contact, there is a joint (or a suture, for adjoining bones of the skull). The bones may be separated by cartilage or fluid. Within the body, there is a range of joint types—ball-and-socket, ellipsoidal, hinge, gliding, pivot, and saddle—each offering a different range of movement. Joints are supported and strengthened by ligaments (tough, white, fibrous, slightly elastic tissues).

Answers

1. Frontal bone, 2. Parietal bone, 3. Temporal bone, 4. Maxilla, 5. Cervical vertebra, 6. Costal cartilage, 7. True rib, 8. Thoracic vertebra, 9. False rib, 10. Lumbar vertebra, 11. Transverse process, 12. Ilium, 13. Sacrum, 14. Coccyx, 15. Ischium, 16. Pubic symphysis, 17. Tarsal bones, 18. Phalanges, 19. Metatarsal bones, 20. Talus, 21. Fibula, 22. Tibia, 23. Patella, 24. Femur, 25. Pubic bone, 26. Metacarpal bones, 27. Phalanges, 28. Carpal bones, 29. Ulna, 30. Radius, 31. Twelfth rib (floating rib), 32. Humerus, 33. Sternum, 34. Clavicle, 35. Mandible, 36. Lower teeth, 37. Upper teeth, 38. Anterior nasal aperture, 39. Orbit

Skeletal System—posterior view

Skeletal System—lateral view

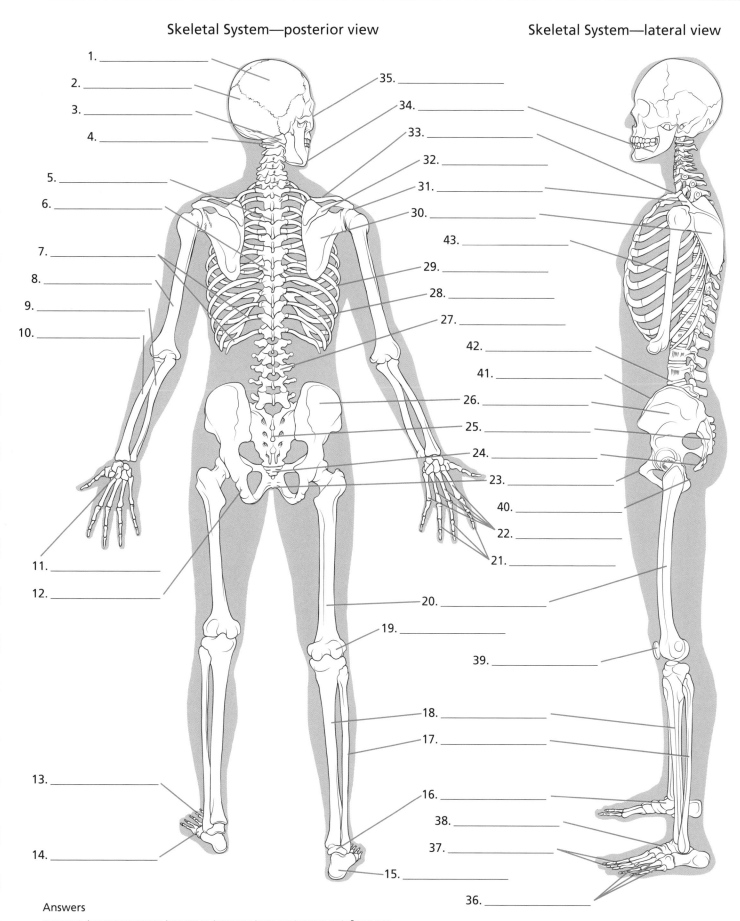

1. _____
2. _____
3. _____
4. _____
5. _____
6. _____
7. _____
8. _____
9. _____
10. _____
11. _____
12. _____
13. _____
14. _____
15. _____

35. _____
34. _____
33. _____
32. _____
31. _____
30. _____
43. _____
29. _____
28. _____
27. _____
42. _____
41. _____
26. _____
25. _____
24. _____
23. _____
40. _____
22. _____
21. _____
20. _____
19. _____
39. _____
18. _____
17. _____
16. _____
38. _____
37. _____
36. _____

Answers

Bone Structure

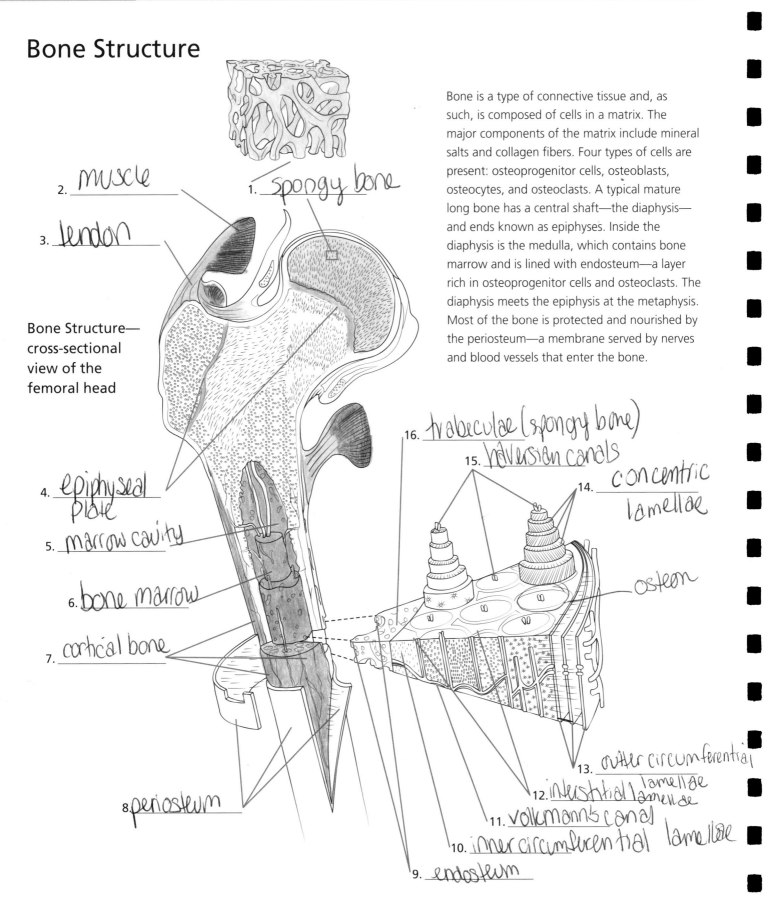

Bone is a type of connective tissue and, as such, is composed of cells in a matrix. The major components of the matrix include mineral salts and collagen fibers. Four types of cells are present: osteoprogenitor cells, osteoblasts, osteocytes, and osteoclasts. A typical mature long bone has a central shaft—the diaphysis—and ends known as epiphyses. Inside the diaphysis is the medulla, which contains bone marrow and is lined with endosteum—a layer rich in osteoprogenitor cells and osteoclasts. The diaphysis meets the epiphysis at the metaphysis. Most of the bone is protected and nourished by the periosteum—a membrane served by nerves and blood vessels that enter the bone.

Bone Structure—
cross-sectional
view of the
femoral head

1. spongy bone
2. muscle
3. tendon
4. epiphyseal plate
5. marrow cavity
6. bone marrow
7. cortical bone
8. periosteum
9. endosteum
10. inner circumferential lamellae
11. volkmann's canal
12. intersitial lamellae
13. outter circumferential lamellae
14. concentric lamellae
15. haversian canals
16. trabeculae (spongy bone)
osteon

Answers

1. Spongy bone, 2. Muscle, 3. Tendon, 4. Epiphyseal plate, 5. Marrow cavity, 6. Bone marrow, 7. Cortical bone, 8. Periosteum, 9. Endosteum, 10. Inner circumferential lamellae, 11. Volkmann's canal, 12. Interstitial lamellae, 13. Outer circumferential lamellae, 14. Concentric lamellae, 15. Haversian canals, 16. Trabecula of spongy bone

Bones of the Skull

The skull forms the skeleton of the head, and consists of the cranium and the mandible. The neurocranium surrounds the brain and part of the brainstem. The facial cranium is the lower part of the skull that underlies the face. Fourteen bones make up the facial cranium, including two nasal bones; two lacrimal bones, which are located in each orbit; the maxillary bones; the mandible; the two palatine bones of the hard palate; the vomer, which, with part of the ethmoid bone, makes up the nasal septum; and the two inferior turbinates of the nose. The rear view of the skull is dominated by the occipital bone in the midline below, with the parietal bones above on each side. The cranial bones overlie the four lobes of the brain. The bones of the skull are linked together by joints known as sutures. Though classed as joints, the connections between the skull bones are fixed and immobile.

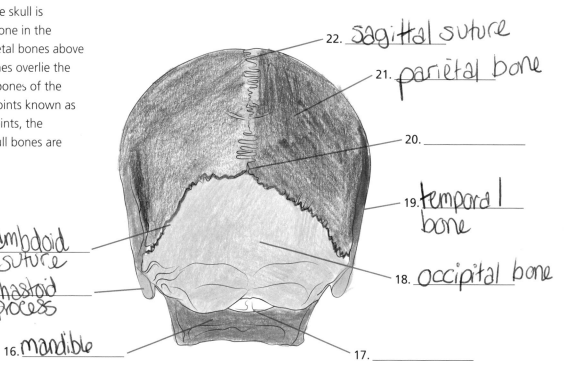

1. parietal bone
13. frontal bone
12. nasal bone
11. _____
10. temporal bone
9. _____
8. zygomatic bone
7. vomer
6. maxilla
5. _____
4. _____
3. mandible
2. Mental foramen

Bones of the Skull—anterior view

22. Sagittal suture
21. parietal bone
20. _____
19. temporal bone
18. occipital bone
14. lambdoid suture
15. Mastoid process
16. mandible
17. _____

Bones of the Skull—posterior view

Answers
1. Parietal bone, 2. Mental foramen, 3. Mandible, 4. Lower (mandibular) teeth, 5. Upper (maxillary) teeth, 6. Maxilla, 7. Nasal septum, 8. Zygomatic bone,
9. Greater wing of sphenoid bone, 10. Temporal bone, 11. Lesser wing of sphenoid bone, 12. Nasal bone, 13. Frontal bone, 14. Lambdoid suture,
15. Mastoid process, 16. Mandible, 17. External occipital protuberance, 18. Occipital bone, 19. Temporal bone, 20. Lambda, 21. Parietal bone, 22. Sagittal suture

Bones of the Skull

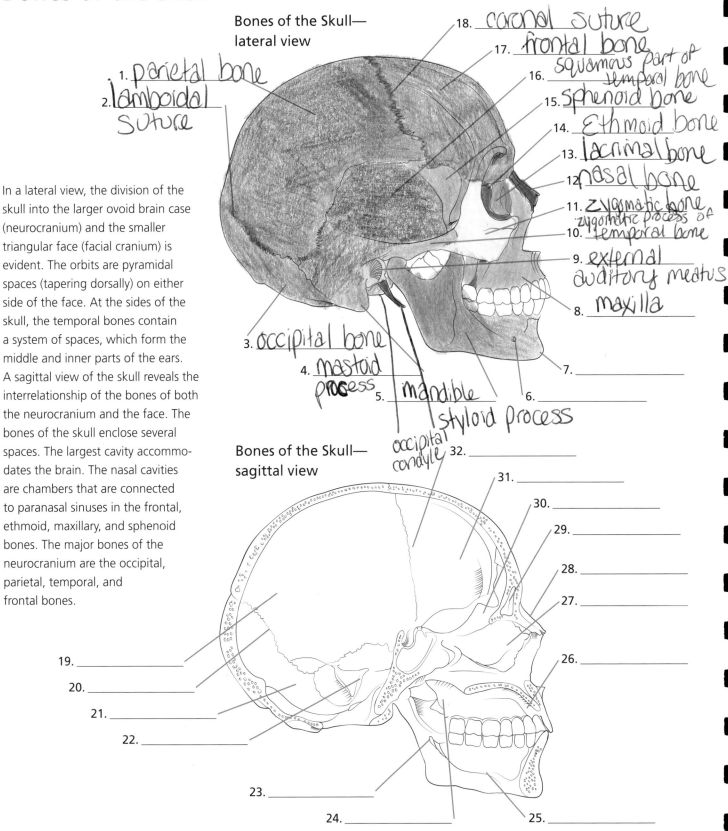

Bones of the Skull— lateral view

In a lateral view, the division of the skull into the larger ovoid brain case (neurocranium) and the smaller triangular face (facial cranium) is evident. The orbits are pyramidal spaces (tapering dorsally) on either side of the face. At the sides of the skull, the temporal bones contain a system of spaces, which form the middle and inner parts of the ears. A sagittal view of the skull reveals the interrelationship of the bones of both the neurocranium and the face. The bones of the skull enclose several spaces. The largest cavity accommodates the brain. The nasal cavities are chambers that are connected to paranasal sinuses in the frontal, ethmoid, maxillary, and sphenoid bones. The major bones of the neurocranium are the occipital, parietal, temporal, and frontal bones.

1. parietal bone
2. lamboidal suture
3. occipital bone
4. mastoid process
5. mandible
6. _____
7. _____
8. maxilla
9. external auditory meatus
10. temporal bone / zygomatic process of
11. zygomatic bone
12. nasal bone
13. lacrimal bone
14. Ethmoid bone
15. sphenoid bone
16. squamous part of temporal bone
17. frontal bone
18. coronal suture

stylord process
occipital condyle

Bones of the Skull— sagittal view

19. _____
20. _____
21. _____
22. _____
23. _____
24. _____
25. _____
26. _____
27. _____
28. _____
29. _____
30. _____
31. _____
32. _____

A number of bones, both neurocranial and facial cranial, come together to form the base of the skull. The base of the skull contains a number of foramina, through which the spinal cord, cranial nerves, and blood vessels enter or exit. The large hole in the occipital bone, the foramen magnum, is for the passage of the spinal cord.

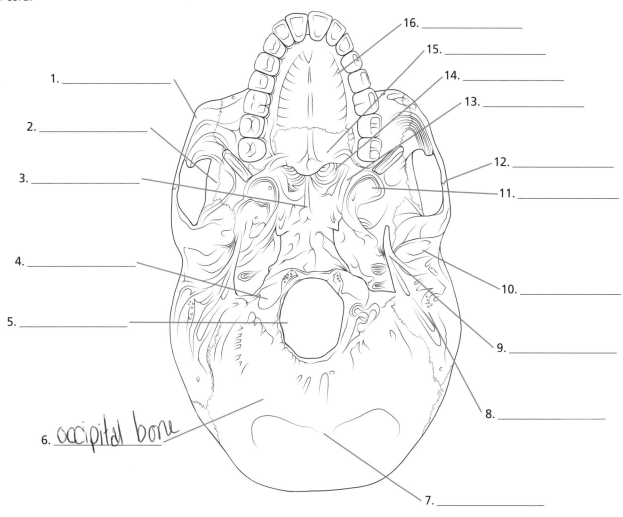

1. _____

2. _____

3. _____

4. _____

5. _____

6. *occipital bone*

7. _____

8. _____

9. _____

10. _____

11. _____

12. _____

13. _____

14. _____

15. _____

16. _____

Base of the Skull—inferior view

Answers

1. Zygomatic bone, 2. Greater wing of sphenoid bone, 3. Vomer, 4. Occipital condyle, 5. Foramen magnum, 6. Occipital bone, 7. External occipital protuberance, 8. Mastoid process, 9. Styloid process, 10. Mandibular fossa, 11. Lateral pterygoid plate, 12. Zygomatic arch, 13. Medial pterygoid plate, 14. Posterior nasal aperture, 15. Palatine bone, 16. Palatine process (maxilla)

Bones of the Head and Face

The jaw consists of two parts: a movable lower jaw, formed by the mandible, and a fixed upper jaw, formed by the maxillary bones. The mandible articulates with the temporal bones on each side at the temporomandibular joints. These joints are reinforced by a capsule and strong ligaments. The teeth are bony structures embedded in the upper and lower jaws. The cavities for the eyes, called the orbits, are formed by the frontal, ethmoid, lacrimal, zygomatic, nasal, palatine, sphenoid, and maxillary bones. The bones that form the outer wall and roof of the orbit are thick and strong, whereas the bones that form the inner walls are thin and fragile.

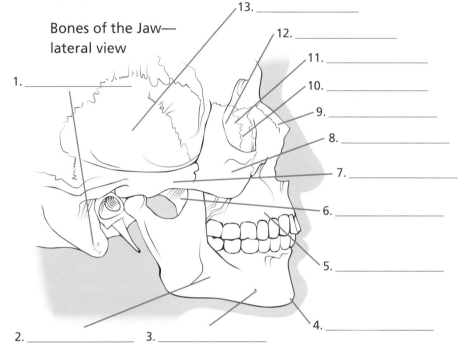

Bones of the Jaw—lateral view

13. _____
12. _____
11. _____
10. _____
9. _____
8. _____
7. _____
6. _____
5. _____
4. _____
1. _____
2. _____
3. _____

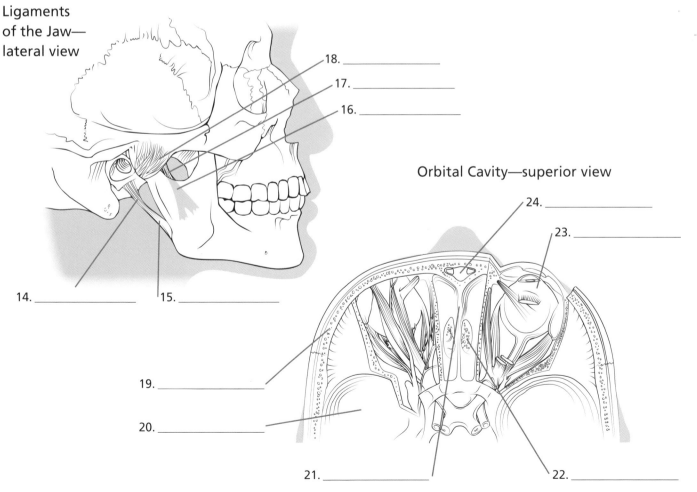

Ligaments of the Jaw—lateral view

18. _____
17. _____
16. _____

14. _____
15. _____

19. _____
20. _____
21. _____

Orbital Cavity—superior view

24. _____
23. _____
22. _____

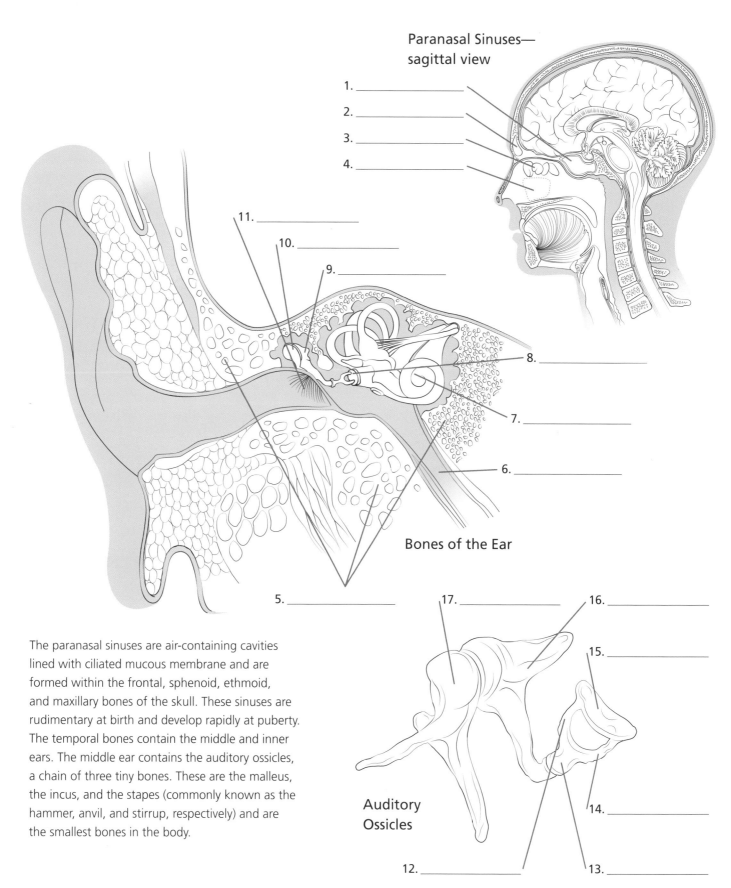

Paranasal Sinuses— sagittal view

1. _____
2. _____
3. _____
4. _____

11. _____
10. _____
9. _____

8. _____
7. _____
6. _____

5. _____

Bones of the Ear

17. _____   16. _____

15. _____

14. _____

Auditory Ossicles

12. _____   13. _____

The paranasal sinuses are air-containing cavities lined with ciliated mucous membrane and are formed within the frontal, sphenoid, ethmoid, and maxillary bones of the skull. These sinuses are rudimentary at birth and develop rapidly at puberty. The temporal bones contain the middle and inner ears. The middle ear contains the auditory ossicles, a chain of three tiny bones. These are the malleus, the incus, and the stapes (commonly known as the hammer, anvil, and stirrup, respectively) and are the smallest bones in the body.

Answers

1. Sphenoidal sinus, 2. Frontal sinus, 3. Ethmoid sinus, 4. Maxillary sinus, 5. Temporal bone, 6. Eustachian (auditory) tube, 7. Cochlea, 8. Stapes, 9. Incus, 10. Malleus, 11. Tympanic membrane (eardrum), 12. Anterior crus of stapes, 13. Stapes, 14. Posterior crus of stapes, 15. Footplate of stapes, 16. Incus, 17. Malleus

Vertebral Column

The vertebral column is a chain of bones divided into five regions—cervical, thoracic, lumbar, sacral, and coccygeal. The cervical region is formed by seven vertebrae, numbered 1–7 from the top down (C1–C7). The thoracic region is formed by twelve vertebrae (T1–T12), all of which have ribs attached to their sides. The lumbar region has five vertebrae (L1–L5). The sacrum is located beneath the lumbar vertebrae and is formed by five fused bones. Four fused bones form the coccyx, which lies at the base of the sacrum. The vertebrae articulate with one another through intervertebral disks.

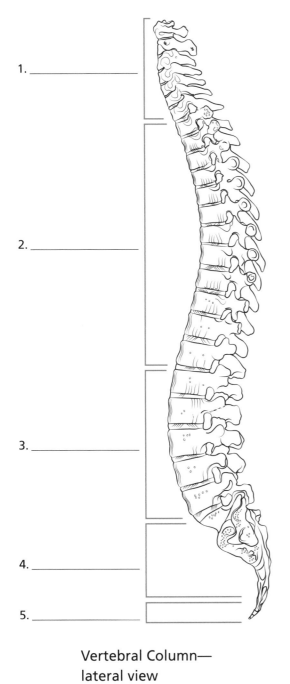

1. _____

2. _____

3. _____

4. _____

5. _____

Vertebral Column—
lateral view

6. _____

7. _____

8. _____

9. _____

10. _____

14. _____

13. _____

12. _____

11. _____

Vertebral Column—
posterior view

Answers

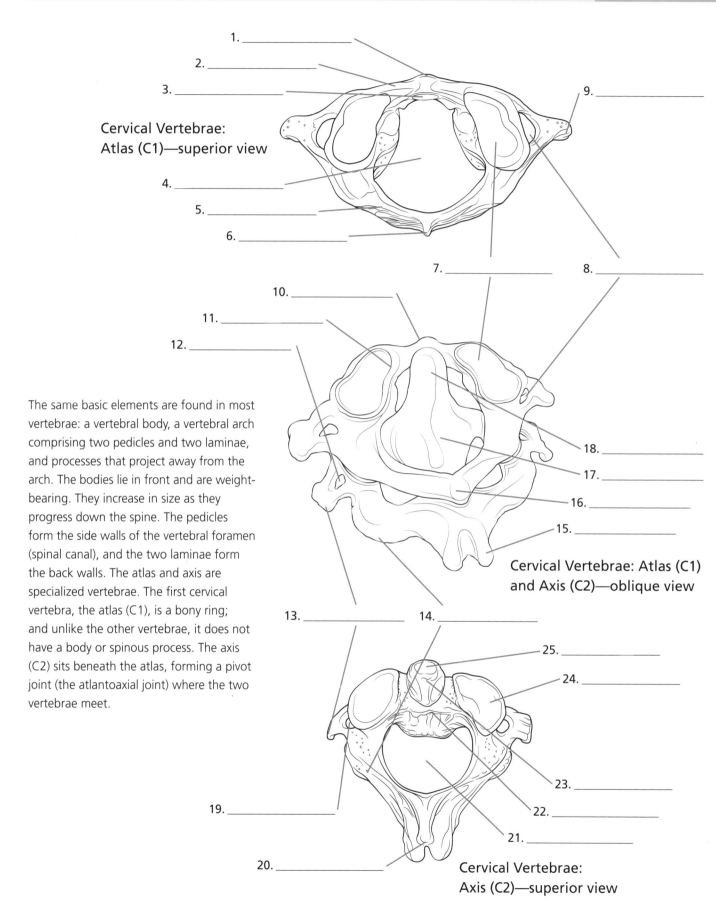

Cervical Vertebrae:
Atlas (C1)—superior view

1. _____
2. _____
3. _____
4. _____
5. _____
6. _____
7. _____
8. _____
9. _____
10. _____
11. _____
12. _____

The same basic elements are found in most vertebrae: a vertebral body, a vertebral arch comprising two pedicles and two laminae, and processes that project away from the arch. The bodies lie in front and are weight-bearing. They increase in size as they progress down the spine. The pedicles form the side walls of the vertebral foramen (spinal canal), and the two laminae form the back walls. The atlas and axis are specialized vertebrae. The first cervical vertebra, the atlas (C1), is a bony ring; and unlike the other vertebrae, it does not have a body or spinous process. The axis (C2) sits beneath the atlas, forming a pivot joint (the atlantoaxial joint) where the two vertebrae meet.

18. _____
17. _____
16. _____
15. _____

Cervical Vertebrae: Atlas (C1)
and Axis (C2)—oblique view

13. _____
14. _____
25. _____
24. _____
23. _____
22. _____
21. _____
19. _____
20. _____

Cervical Vertebrae:
Axis (C2)—superior view

Answers

1. Anterior tubercle, 2. Anterior arch, 3. Articular facet for dens, 4. Vertebral foramen, 5. Posterior arch, 6. Posterior tubercle, 7. Superior articular facet, 8. Transverse foramen, 9. Transverse process, 10. Anterior tubercle of atlas, 11. Anterior arch of atlas, 12. Transverse foramen of axis, 13. Transverse process of axis, 14. Inferior articular process, 15. Spinous process of axis, 16. Posterior tubercle of atlas, 17. Body of axis, 18. Dens of axis, 19. Transverse foramen, 20. Spinous process, 21. Vertebral foramen, 22. Body of axis, 23. Dens of axis, 24. Superior articular facet, 25. Facet for atlas

Vertebral Column

superior view

Typical
Cervical Vertebra

lateral view

1. _____

2. _____

3. _____

4. _____

9. _____

8. _____

7. _____

13. _____

6. _____

5. _____

10. _____

11. _____

12. _____

There are seven cervical vertebrae (C1–C7). C1 and C2 are known as the atlas and axis. The remaining cervical vertebrae are similar in structure to most other vertebrae and have a body, a vertebral arch comprising two pedicles and two laminae, and transverse processes that project away from the arch. Lying between the cervical and lumbar regions, the thoracic region of the vertebral column is formed by twelve vertebrae (T1–T12), all of which have ribs attached to their sides. The lumbar (lower back) region of the vertebral column has five vertebrae (L1–L5). The lumbar vertebrae are the largest of the vertebrae.

superior view

Thoracic Vertebra

lateral view

20. _____

19. _____

18. _____

17. _____

16. _____

15. _____

14. _____

22. _____

21. _____

superior view

Lumbar Vertebra

lateral view

30. _____

29. _____

28. _____

27. _____

26. _____

23. _____

24. _____

25. _____

32. _____

31. _____

Answers

The intervertebral disks are flexible cartilaginous structures that lie between adjacent vertebrae. Each disk has two parts: a central mass (the nucleus pulposus) and a surrounding fibrous layer (the annulus fibrosus). The thickness and shape of the disks vary in different parts of the vertebral column—they are relatively thin and flat in the thoracic region, thicker in the cervical region, and wedge-shaped in the lumbar region. The sacrum lies at the lower end of the spine beneath the lumbar vertebrae and forms part of the bony pelvis. A single curved bone in the adult, the sacrum develops from five separate vertebrae (S1–S5), which fuse to each other during early development to form a single bone. The coccyx, or tailbone, lies at the base of the spine. It is formed by the fusion of four vertebrae—a process that is usually complete by the time a person reaches their late twenties.

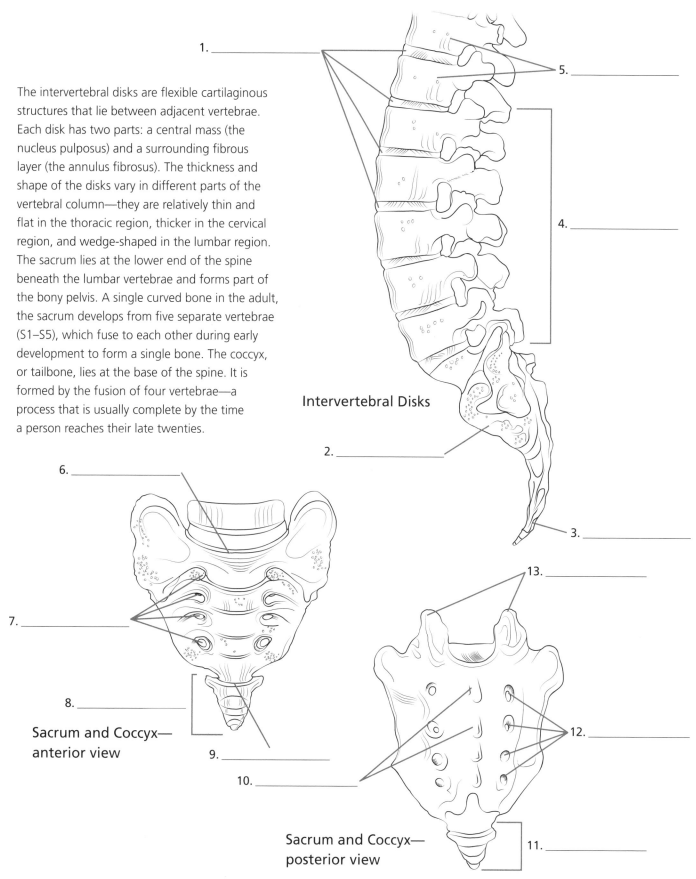

1. _____

5. _____

4. _____

Intervertebral Disks

2. _____

3. _____

6. _____

7. _____

8. _____

9. _____

10. _____

Sacrum and Coccyx— anterior view

13. _____

12. _____

11. _____

Sacrum and Coccyx— posterior view

Rib Cage and Clavicle

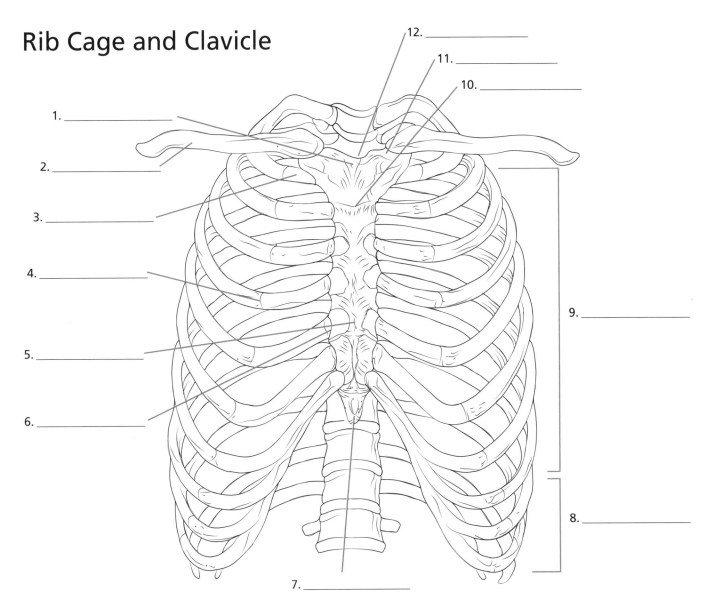

1. _____

2. _____

3. _____

4. _____

5. _____

6. _____

7. _____

8. _____

9. _____

10. _____

11. _____

12. _____

Rib Cage—anterior view

There are twelve pairs of ribs. Pairs 1–7 are known as "true ribs," and these ribs span around from the thoracic vertebrae at the back to the front, where they are joined to the sternum by costal cartilage. Pairs 8–10 are known as "false ribs," and they span around from the vertebrae at the back but do not meet up with the sternum independently; rather, they are joined by costal cartilages to the seventh costal cartilage. Pairs 11–12 are known as "floating ribs" because they do not attach to the sternum at all, either directly or indirectly. These two pairs of ribs do not extend fully around to the front of the chest. The clavicles are a pair of short horizontal bones above the rib cage and are attached to the sternum and to the two scapulas on either side.

Answers

1. Manubrium of sternum, 2. Clavicle, 3. Costal cartilage, 4. Costochondral joint, 5. Body of sternum, 6. Sternocostal joint, 7. Xiphoid process, 8. False ribs (pairs 8–10), 9. True ribs (pairs 1–7), 10. Sternal angle, 11. Sternoclavicular joint, 12. Suprasternal notch

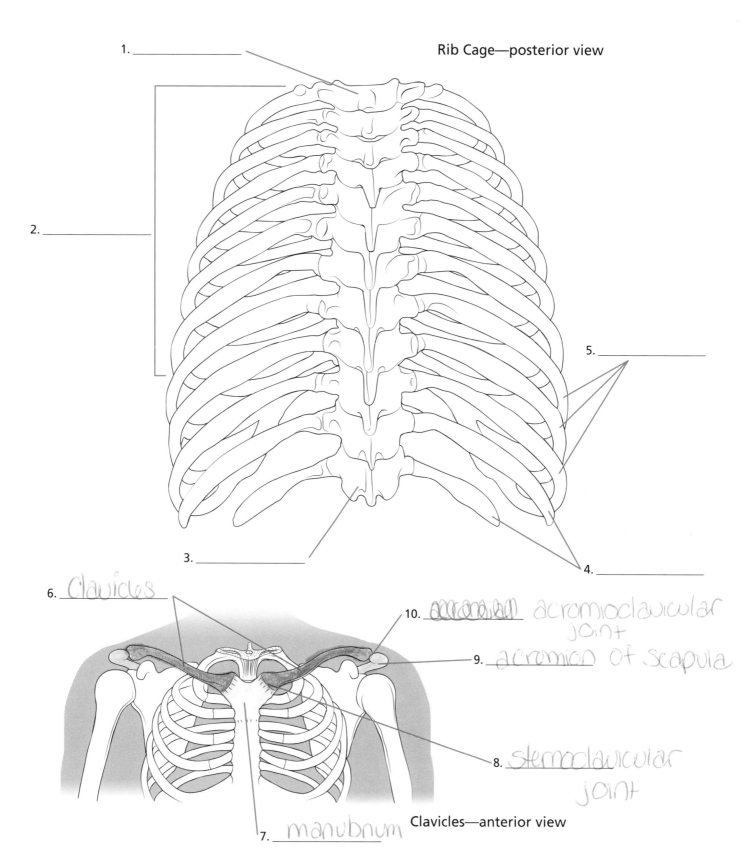

Rib Cage—posterior view

1. _____

2. _____

3. _____

4. _____

5. _____

6. Clavicles

7. manubrum

8. Sternoclavicular joint

9. acromion of scapula

10. acromial acromioclavicular joint

Clavicles—anterior view

Answers

Bones of the Upper Limb

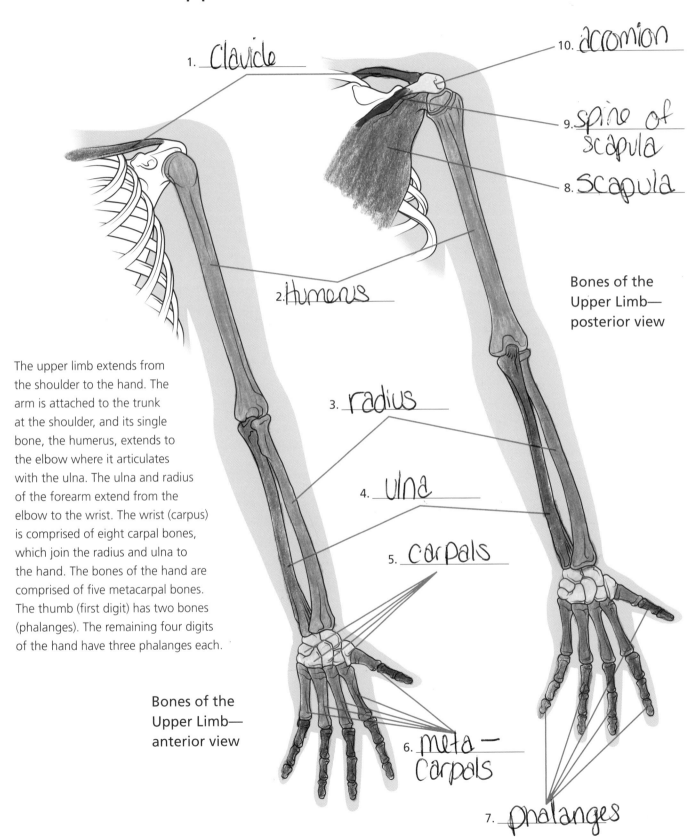

1. Clavicle

10. acromion

9. spine of scapula

8. scapula

2. Humerus

Bones of the
Upper Limb—
posterior view

3. radius

4. ulna

5. carpals

The upper limb extends from the shoulder to the hand. The arm is attached to the trunk at the shoulder, and its single bone, the humerus, extends to the elbow where it articulates with the ulna. The ulna and radius of the forearm extend from the elbow to the wrist. The wrist (carpus) is comprised of eight carpal bones, which join the radius and ulna to the hand. The bones of the hand are comprised of five metacarpal bones. The thumb (first digit) has two bones (phalanges). The remaining four digits of the hand have three phalanges each.

Bones of the
Upper Limb—
anterior view

6. meta-carpals

7. phalanges

The humerus of the arm and the scapula are the two major components of the shoulder joint, with the clavicle providing stability to the joint. The clavicle and scapula form the pectoral girdle, responsible for the attachment of the upper limb to the trunk. The clavicle acts as a strut to hold the upper limb away from the center of the body. The scapula is a flat triangle-shaped bone that covers part of the upper back. Together, the scapula and humerus form the highly mobile shoulder joint. A multiaxial ball-and-socket joint, the shoulder joint is capable of movement in almost any direction.

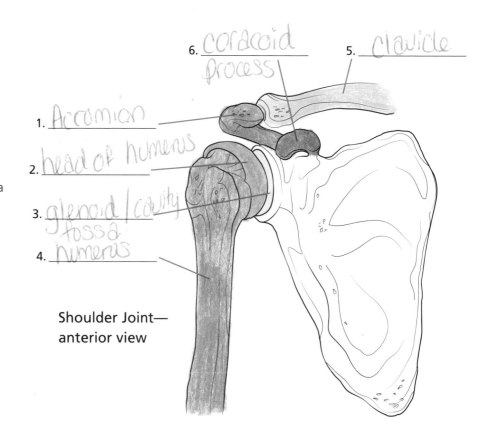

6. coracoid process
5. clavicle
1. Acromion
2. head of humerus
3. glenoid/cavity fossa
4. humerus

Shoulder Joint— anterior view

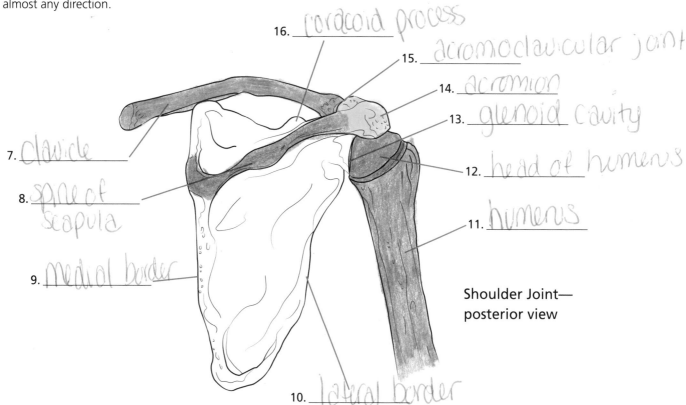

16. coracoid process
15. acromoclavicular joint
14. acromion
13. glenoid cavity
12. head of humerus
11. humerus
7. Clavicle
8. spine of scapula
9. Medial border
10. lateral border

Shoulder Joint— posterior view

Answers

Bones of the Upper Limb

Shoulder Joint—superior view

1. _____

2. _____

3. _____

The extremely mobile ball-and-socket joint of the shoulder is formed where the round ball of the humerus and the shallow surface of the glenoid cavity of the scapula meet. A number of strong ligaments keep the shoulder joint stable, including the glenohumeral ligaments, which link the humerus and scapula, and the coracohumeral ligament, which links the humerus and the coracoid process. Adding extra stability to the region are the coracoacromial ligament, which joins the coracoid process and the acromion; the acromioclavicular ligament, which joins the acromion to the clavicle; and the coracoclavicular ligament, which joins the coracoid process and the clavicle. The transverse humeral ligament spans the two tubercles at the head of the humerus, acting to hold firm the tendons of biceps brachii.

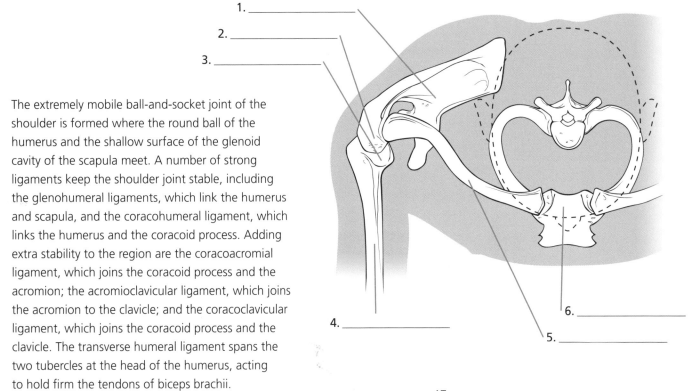

4. _____

5. _____

6. _____

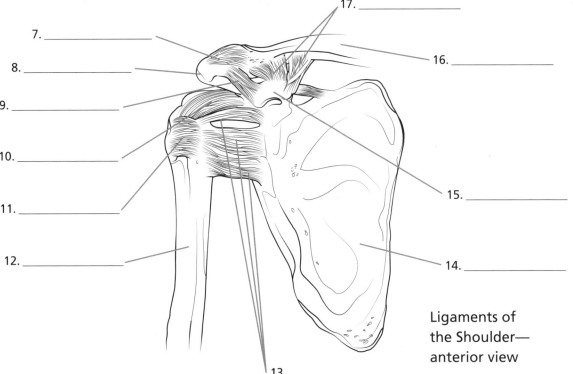

7. _____

8. _____

9. _____

10. _____

11. _____

12. _____

13. _____

14. _____

15. _____

16. _____

17. _____

Ligaments of the Shoulder— anterior view

The humerus of the arm consists of a long cylindrical shaft with a rounded head at its proximal end, which articulates with the scapula at its glenoid fossa to form the shoulder joint, and with paired condyles at its distal end, which articulate with the radius and ulna to form the elbow joint. Other landmarks of the humerus include the deltoid tuberosity, the medial and lateral epicondyles, and the trochlea.

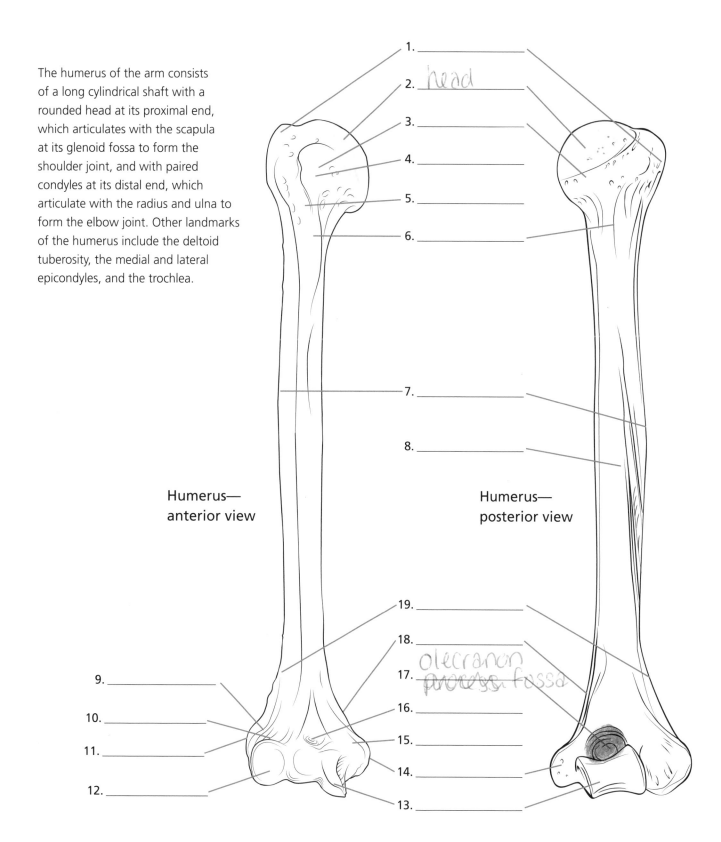

1. _____

2. head

3. _____

4. _____

5. _____

6. _____

7. _____

8. _____

Humerus— anterior view

Humerus— posterior view

19. _____

18. _____

17. olecranon process fossa

16. _____

15. _____

14. _____

13. _____

9. _____

10. _____

11. _____

12. _____

Bones of the Upper Limb

The elbow is the joint between the expanded distal end of the humerus of the arm and the proximal ends of the ulna and radius of the forearm. This is a hinge joint, allowing flexion (bending) and extension (straightening). The principal flexion–extension action occurs between the distal end of the humerus and the proximal end of the ulna. The articulation between the radius and ulna forms a pivot joint, which allows rotational movement of the radius around the ulna. The bones of the elbow joint (humerus, radius, and ulna) are held together by strong fibrous ligaments. Joining the humerus to the ulna and olecranon, on the medial (inner) side of the joint, is the ulnar collateral ligament. On the lateral side of the joint, the radial collateral ligament joins the humerus to the radius, affixing to the radius at the annular ligament, which joins the head of the radius to the ulna.

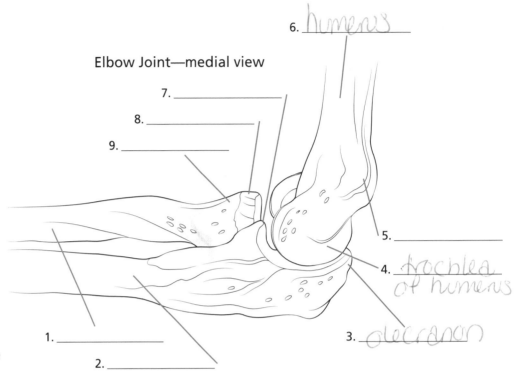

Elbow Joint—medial view

6. _humerus_

7. _____

8. _____

9. _____

5. _____

4. _trochlea of humerus_

3. _olecranon_

1. _____

2. _____

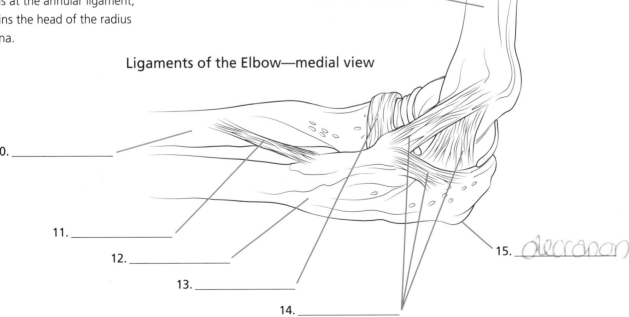

Ligaments of the Elbow—medial view

16. _____

10. _____

11. _____

12. _____

13. _____

14. _____

15. _olecranon_

Answers

The radius is one of the two bones of the forearm. It is located on the thumb side, lying parallel to and rotating around the ulna. Near the uppermost end of the radius is a raised and roughened area called the radial tuberosity. At its larger distal end, the radius forms part of the wrist. The ulna lies on the medial side of the forearm, extending from the elbow to the wrist. The ulna is a long bone of irregular cross section, thickest at the proximal end and tapering toward the distal end. It projects above and behind the elbow.

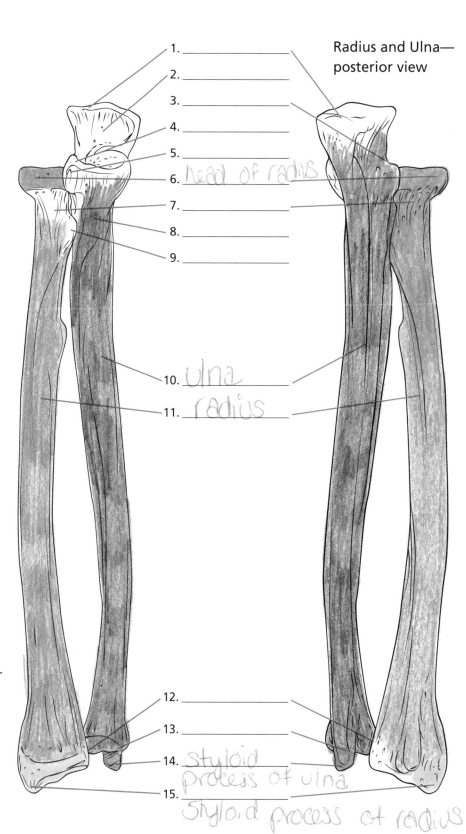

Radius and Ulna— posterior view

1. _____
2. _____
3. _____
4. _____
5. _____
6. _head of radius_
7. _____
8. _____
9. _____

10. _ulna_
11. _radius_

Radius and Ulna— anterior view

12. _____
13. _____
14. _styloid process of ulna_
15. _styloid process of radius_

Answers

Bones of the Wrist and Hand

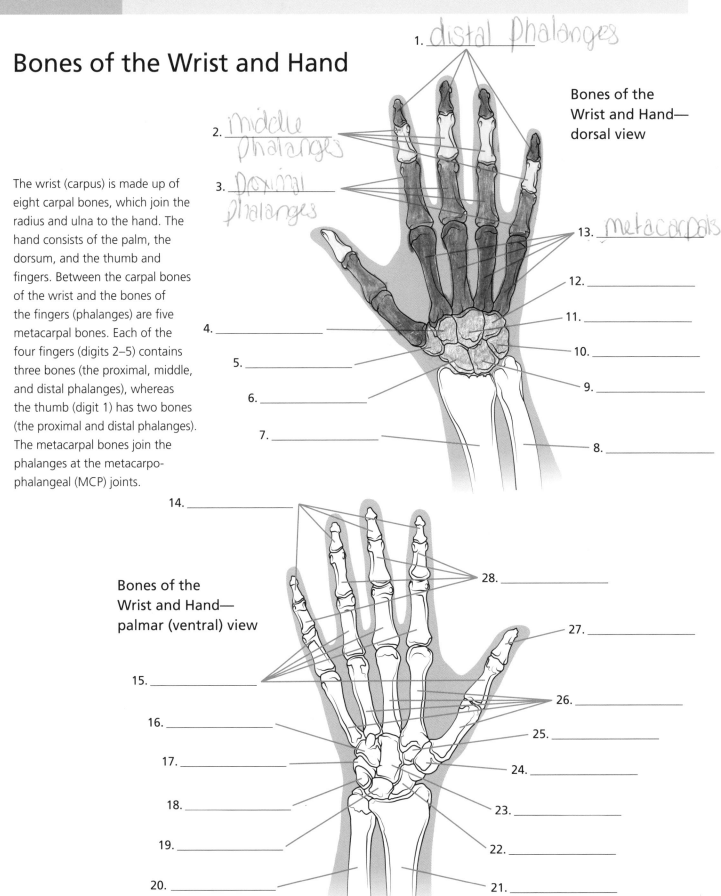

Bones of the Wrist and Hand— dorsal view

The wrist (carpus) is made up of eight carpal bones, which join the radius and ulna to the hand. The hand consists of the palm, the dorsum, and the thumb and fingers. Between the carpal bones of the wrist and the bones of the fingers (phalanges) are five metacarpal bones. Each of the four fingers (digits 2–5) contains three bones (the proximal, middle, and distal phalanges), whereas the thumb (digit 1) has two bones (the proximal and distal phalanges). The metacarpal bones join the phalanges at the metacarpophalangeal (MCP) joints.

1. distal phalanges
2. middle phalanges
3. proximal phalanges
13. metacarpals
12. _____
11. _____
10. _____
9. _____
8. _____
7. _____
6. _____
5. _____
4. _____

Bones of the Wrist and Hand— palmar (ventral) view

14. _____
28. _____
27. _____
26. _____
25. _____
24. _____
23. _____
22. _____
21. _____
20. _____
19. _____
18. _____
17. _____
16. _____
15. _____

Answers

1. Distal phalanges, 2. Middle phalanges, 3. Proximal phalanges, 4. Proximal phalanges, 5. Trapezoid, 6. Trapezium, 7. Scaphoid, 8. Ulna, 9. Lunate, 10. Triquetral,
11. Capitate, 12. Hamate, 13. Metacarpal bones, 14. Distal phalanges, 15. Proximal phalanges, 16. Hamate, 17. Triquetral, 18. Pisiform, 19. Lunate, 20. Ulna,
21. Radius, 22. Scaphoid, 23. Capitate, 24. Trapezium, 25. Trapezoid, 26. Metacarpal bones, 27. Distal phalanx of thumb, 28. Middle phalanges

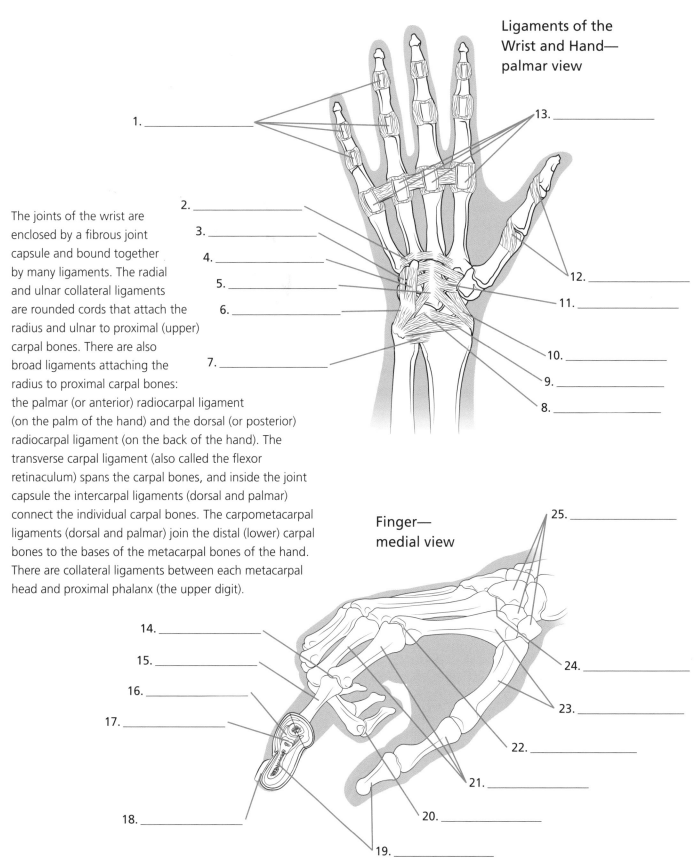

Ligaments of the Wrist and Hand—palmar view

1. _____

2. _____

3. _____

4. _____

5. _____

6. _____

7. _____

8. _____

9. _____

10. _____

11. _____

12. _____

13. _____

The joints of the wrist are enclosed by a fibrous joint capsule and bound together by many ligaments. The radial and ulnar collateral ligaments are rounded cords that attach the radius and ulnar to proximal (upper) carpal bones. There are also broad ligaments attaching the radius to proximal carpal bones: the palmar (or anterior) radiocarpal ligament (on the palm of the hand) and the dorsal (or posterior) radiocarpal ligament (on the back of the hand). The transverse carpal ligament (also called the flexor retinaculum) spans the carpal bones, and inside the joint capsule the intercarpal ligaments (dorsal and palmar) connect the individual carpal bones. The carpometacarpal ligaments (dorsal and palmar) join the distal (lower) carpal bones to the bases of the metacarpal bones of the hand. There are collateral ligaments between each metacarpal head and proximal phalanx (the upper digit).

Finger—medial view

14. _____

15. _____

16. _____

17. _____

18. _____

19. _____

20. _____

21. _____

22. _____

23. _____

24. _____

25. _____

Answers

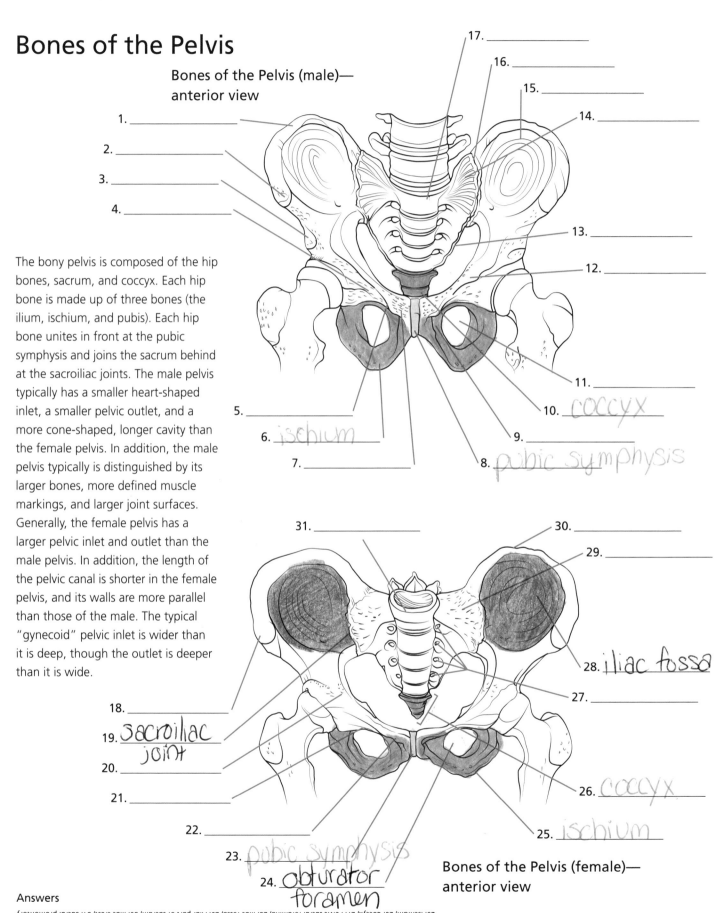

Bones of the Pelvis

Bones of the Pelvis (male)— anterior view

1. _____

2. _____

3. _____

4. _____

The bony pelvis is composed of the hip bones, sacrum, and coccyx. Each hip bone is made up of three bones (the ilium, ischium, and pubis). Each hip bone unites in front at the pubic symphysis and joins the sacrum behind at the sacroiliac joints. The male pelvis typically has a smaller heart-shaped inlet, a smaller pelvic outlet, and a more cone-shaped, longer cavity than the female pelvis. In addition, the male pelvis typically is distinguished by its larger bones, more defined muscle markings, and larger joint surfaces. Generally, the female pelvis has a larger pelvic inlet and outlet than the male pelvis. In addition, the length of the pelvic canal is shorter in the female pelvis, and its walls are more parallel than those of the male. The typical "gynecoid" pelvic inlet is wider than it is deep, though the outlet is deeper than it is wide.

17. _____

16. _____

15. _____

14. _____

13. _____

12. _____

11. _____

10. coccyx

9. _____

8. pubic symphysis

5. _____

6. ischium

7. _____

18. _____

19. sacroiliac joint

20. _____

21. _____

22. _____

23. pubic symphysis

24. obturator foramen

31. _____

30. _____

29. _____

28. iliac fossa

27. _____

26. coccyx

25. ischium

Bones of the Pelvis (female)— anterior view

Answers

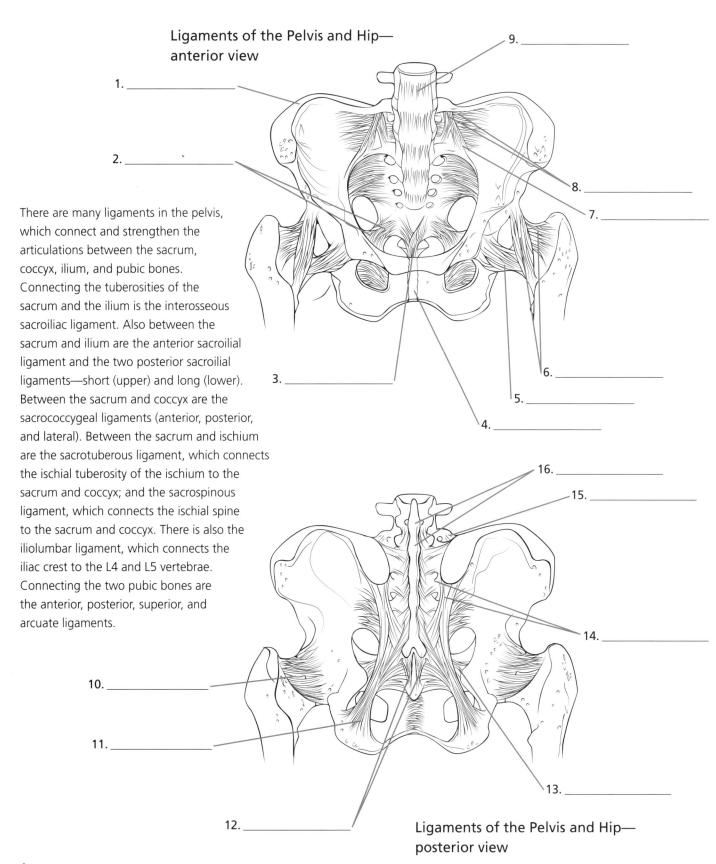

Ligaments of the Pelvis and Hip— anterior view

1. _____

2. _____

9. _____

8. _____

7. _____

6. _____

5. _____

4. _____

3. _____

There are many ligaments in the pelvis, which connect and strengthen the articulations between the sacrum, coccyx, ilium, and pubic bones. Connecting the tuberosities of the sacrum and the ilium is the interosseous sacroiliac ligament. Also between the sacrum and ilium are the anterior sacroilial ligament and the two posterior sacroilial ligaments—short (upper) and long (lower). Between the sacrum and coccyx are the sacrococcygeal ligaments (anterior, posterior, and lateral). Between the sacrum and ischium are the sacrotuberous ligament, which connects the ischial tuberosity of the ischium to the sacrum and coccyx; and the sacrospinous ligament, which connects the ischial spine to the sacrum and coccyx. There is also the iliolumbar ligament, which connects the iliac crest to the L4 and L5 vertebrae. Connecting the two pubic bones are the anterior, posterior, superior, and arcuate ligaments.

16. _____

15. _____

14. _____

10. _____

11. _____

13. _____

12. _____

Ligaments of the Pelvis and Hip— posterior view

Answers

Bones of the Lower Limb

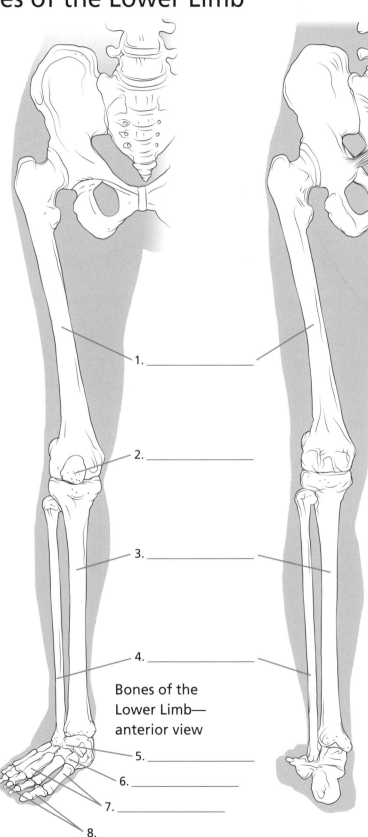

1. _____

2. _____

3. _____

4. _____

Bones of the Lower Limb— anterior view

5. _____

6. _____

7. _____

8. _____

Bones of the Lower Limb— posterior view

The bones of the lower limb include the femur in the thigh and the tibia and fibula in the leg. The femur is the longest bone in the body. The femur articulates with the kneecap (patella), a bone within the tendon of the quadriceps muscle. The tibia is the second longest bone in the body. The tibia and fibula together form the bones of the leg. The fibula is thin and bears little weight. The tibia and fibula articulate with the talus bone at the ankle joint. The bones of the foot are the tarsals, metatarsals, and phalanges.

Answers

1. Femur, 2. Patella, 3. Tibia, 4. Fibula, 5. Talus, 6. Tarsal bones, 7. Metatarsal bones, 8. Phalanges

Extending from the hip to the knee, the femur (thigh bone) is the longest and strongest bone in the body. Features of the femur include a rounded head, a long neck with two enlargements (the greater trochanter and the lesser trochanter), and two pro-tuberances at the distal end of the femur (the lateral condyle and the medial condyle).

1. _____

2. _____

3. _____

4. _____

5. _____

6. _____

7. _____

8. _____

Femur— anterior view

Femur— posterior view

9. _____

10. _____

11. _____

12. _____

13. _____

15. _____

16. _____

14. _____

Answers

Bones of the Lower Limb

Ligaments play an important role in strengthening the knee joint. The medial side is reinforced by the tibial (medial) collateral ligament. The lateral side is reinforced by the fibular (lateral) collateral ligament. The two cruciate ligaments are internal to the joint capsule. The knee joint is a complex hinge joint between the femur, tibia, and patella. The lower end of the femur has a concave surface at the front, into which the back of the patella fits, and two rounded condyles at its base. The upper surface of the tibia is relatively flat. Depth to the tibial plateaus is provided by the medial and lateral menisci, which are wedge-shaped fibrocartilages. Each of the rounded condyles of the femur fits into the shallow sockets formed by the corresponding tibial plateau bounded by the menisci. The articulating surfaces are lined with hyaline cartilage.

1. _____

2. _____

3. _____

4. _____

5. _____

6. _____

Bones and Ligaments of the Knee— anterior view

12. _____

11. _____

10. _____

9. _____

8. _____

7. _____

13. _____

14. _____

15. _____

16. _____

19. _____

18. _____

17. _____

Bones of the Knee— lateral view

Answers

The tibia (commonly known as the shinbone) is the innermost bone of the leg and the thicker of the two bones. The second longest bone in the body, the tibia features a broad head and a cylindrical shaft that widens at the lower end to include the medial malleolus. The medial malleolus of the tibia is the small protruding bump on the inside of the ankle. The fibula is the long slender bone on the outside of the leg, extending from just below the knee to the ankle, where its lower end forms the outer side of the ankle joint.

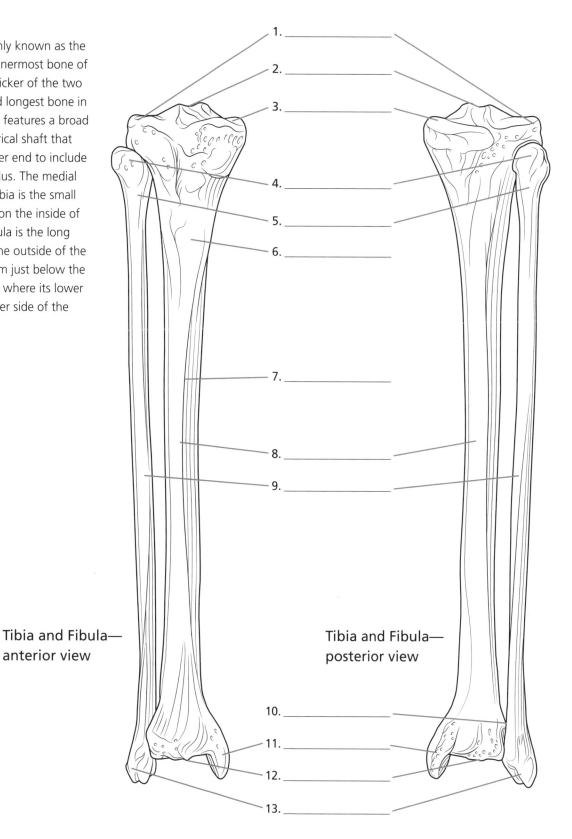

1. _____
2. _____
3. _____
4. _____
5. _____
6. _____
7. _____
8. _____
9. _____

Tibia and Fibula—
anterior view

Tibia and Fibula—
posterior view

10. _____
11. _____
12. _____
13. _____

Answers

1. Lateral tibial condyle, 2. Intercondylar eminence, 3. Medial tibial condyle, 4. Head of fibula, 5. Neck of fibula, 6. Tibial tuberosity, 7. Anterior border, 8. Tibia, 9. Fibula, 10. Fibular notch, 11. Medial malleolus, 12. Inferior articular surface, 13. Lateral malleolus

Bones of the Ankle and Foot

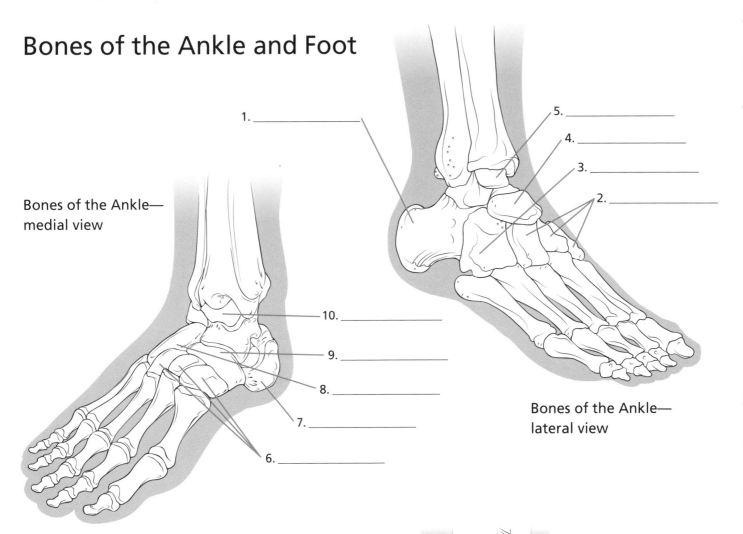

Bones of the Ankle—
medial view

1. _____

10. _____

9. _____

8. _____

7. _____

6. _____

5. _____

4. _____

3. _____

2. _____

Bones of the Ankle—
lateral view

The ankle attaches the tibia and fibula of the leg to the talus of the foot. Prominent features of the ankle joint include the medial malleolus on the tibia and the lateral malleolus on the fibula. The two malleoli, together with part of the tibia, form a socket in which the talus can move. The bones of the foot are the tarsals (seven bones in total: the talus, calcaneus, navicular, cuboid, and the three cuneiform bones [lateral, intermediate, and medial]); the five metatarsals; and the fourteen phalanges of the toes. The tarsus (the back half of the foot) is formed by the seven irregularly shaped tarsal bones.

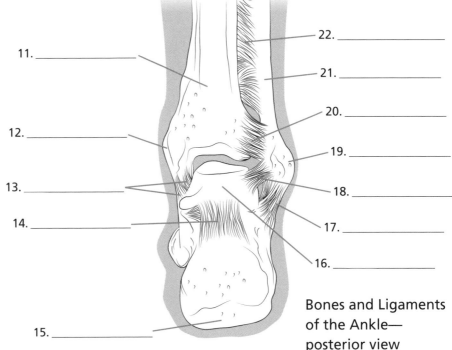

11. _____

12. _____

13. _____

14. _____

15. _____

22. _____

21. _____

20. _____

19. _____

18. _____

17. _____

16. _____

Bones and Ligaments
of the Ankle—
posterior view

Answers

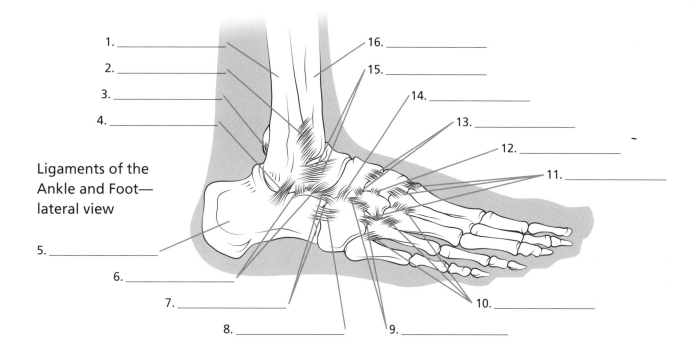

Ligaments of the Ankle and Foot— lateral view

1. _____
2. _____
3. _____
4. _____
5. _____
6. _____
7. _____
8. _____
9. _____
10. _____
11. _____
12. _____
13. _____
14. _____
15. _____
16. _____

Numerous ligaments make the ankle joint strong and stable. On the lateral side of the joint, three cordlike ligaments (the anterior talofibular ligament, the posterior talofibular ligament, and the calcaneofibular ligament) attach the lateral malleolus of the fibula to the talus and calcaneus. On the medial side of the joint, the broad, strong, triangular deltoid (medial) ligament connects the medial malleolus of the tibia to three of the tarsal bones (the talus, the navicular, and the calcaneus). Ligaments also strengthen the joints between each of the bones of the talus, and small ligaments attach the talus bones to adjoining metatarsal bones.

Ligaments of the Ankle and Foot— medial view

17. _____
18. _____
19. _____
20. _____
21. _____
22. _____
23. _____
24. _____
25. _____
26. _____
27. _____
28. _____
29. _____

Answers

1. Fibula, 2. Anterior tibiofibular ligament, 3. Posterior tibiofibular ligament, 4. Calcaneofibular ligament, 5. Calcaneus, 6. Talocalcaneal ligament, 7. Bifurcate ligament, 8. Dorsal calcaneocuboid ligament, 9. Dorsal cuneocuboid ligament, 10. Dorsal cuboideonavicular ligament, 11. Dorsal metatarsal ligaments, 12. Dorsal intercuneiform ligament, 13. Dorsal tarsometatarsal ligaments, 14. Dorsal cuboideonavicular ligament, 15. Anterior talofibular ligament, 16. Tibia, 17. Tibiocalcaneal part of medial deltoid ligament, 18. Anterior tibiotalar part of medial deltoid ligament, 19. Dorsal talonavicular ligament, 20. Dorsal tarsometatarsal ligaments, 21. Dorsal intercuneiform ligament, 22. Dorsal cuneonavicular ligaments, 23. Tibionavicular part of medial deltoid ligament, 24. Short plantar ligament, 25. Long plantar ligament, 26. Plantar calcaneonavicular (spring) ligament, 27. Posterior talocalcaneal ligament, 28. Medial talocalcaneal ligament, 29. Posterior tibiotalar part of medial deltoid ligament

Body Movements

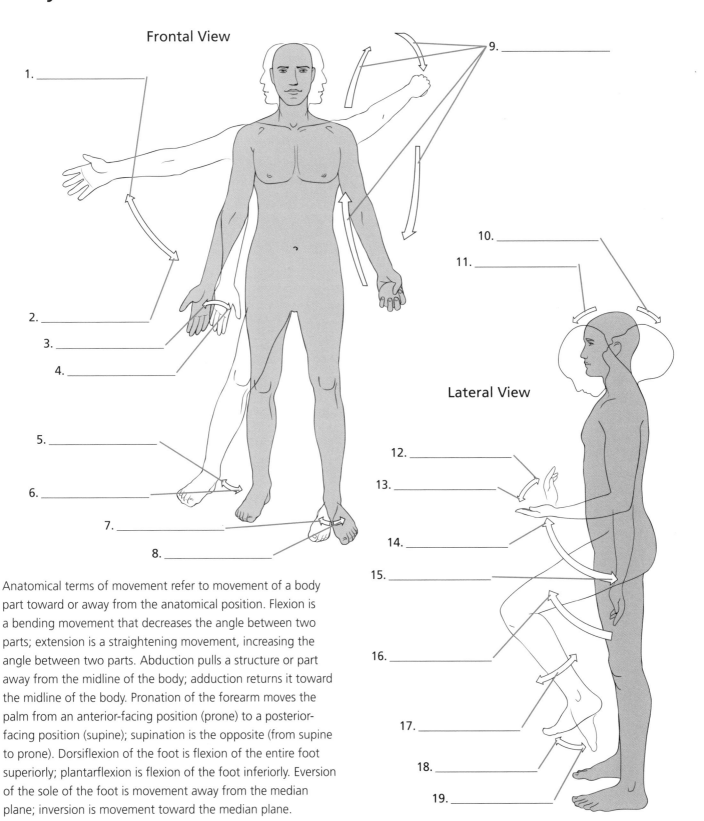

Frontal View

1. _____
2. _____
3. _____
4. _____
5. _____
6. _____
7. _____
8. _____
9. _____

Lateral View

10. _____
11. _____
12. _____
13. _____
14. _____
15. _____
16. _____
17. _____
18. _____
19. _____

Anatomical terms of movement refer to movement of a body part toward or away from the anatomical position. Flexion is a bending movement that decreases the angle between two parts; extension is a straightening movement, increasing the angle between two parts. Abduction pulls a structure or part away from the midline of the body; adduction returns it toward the midline of the body. Pronation of the forearm moves the palm from an anterior-facing position (prone) to a posterior-facing position (supine); supination is the opposite (from supine to prone). Dorsiflexion of the foot is flexion of the entire foot superiorly; plantarflexion is flexion of the foot inferiorly. Eversion of the sole of the foot is movement away from the median plane; inversion is movement toward the median plane.

Answers

1. Abduction, 2. Adduction, 3. Supination, 4. Pronation, 5. Abduction, 6. Adduction, 7. Inversion, 8. Eversion, 9. Circumduction, 10. Extension, 11. Flexion, 12. Wrist flexion, 13. Wrist extension, 14. Elbow flexion, 15. Elbow extension, 16. Flexion, 17. Medial and lateral rotation, 18. Dorsiflexion, 19. Plantarflexion

Joints

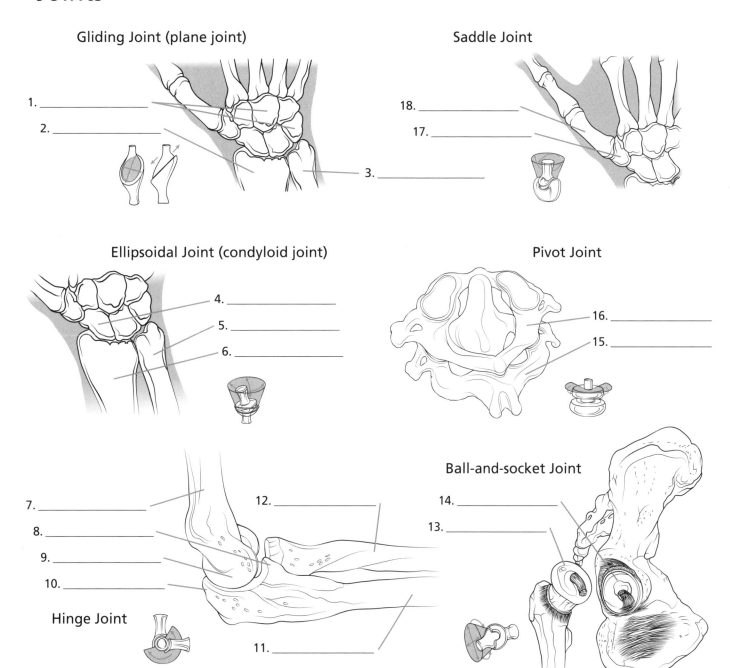

Gliding Joint (plane joint)

1. _____
2. _____
3. _____

Saddle Joint

18. _____
17. _____

Ellipsoidal Joint (condyloid joint)

4. _____
5. _____
6. _____

Pivot Joint

16. _____
15. _____

7. _____
8. _____
9. _____
10. _____

Hinge Joint

12. _____
11. _____

Ball-and-socket Joint

14. _____
13. _____

There are six types of synovial joints found in the appendicular skeleton: ball-and-socket, ellipsoidal (condyloid), gliding (plane), hinge, pivot, and saddle joints. Each permits a different range of movement, with the ball-and-socket joint being the most mobile. In a synovial joint, the ends of the bones are smooth and are covered by articular cartilage with an extremely low coefficient of friction. The two bones are bound together by a capsule of fibrous tissue. The fibrous capsule is lined on the inside with a synovial membrane that secretes synovial fluid to lubricate the joint and nourish the cartilage. The joint is reinforced by ligaments, which are made up of fibers of connective tissue.

Answers

1. Carpal bones, 2. Radius, 3. Ulna, 4. Scaphoid bone, 5. Ulna, 6. Radius, 7. Humerus, 8. Coronoid process of ulna, 9. Trochlea (of humerus), 10. Olecranon, 11. Ulna, 12. Radius, 13. Head of femur (ball), 14. Acetabular fossa (socket), 15. Axis, 16. Atlas, 17. Trapezium bone, 18. Metacarpal bone of thumb

Joints

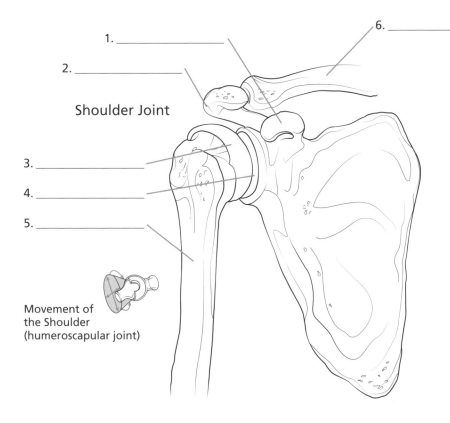

1. _____
2. _____

Shoulder Joint

6. _____

3. _____
4. _____
5. _____

Movement of
the Shoulder
(humeroscapular joint)

The shoulder joint is a multiaxial ball-and-socket joint between the humerus and the scapula, allowing movement in almost any direction. The humerus can be flexed, extended, abducted, adducted, and rotated around its axis. The temporomandibular joint is the joint between the mandible and the skull. This joint allows gliding (backward, forward, and sideways) movements as well as a hinge movement. The elbow is the joint between the humerus and the radius and ulna. The ulna articulates with the humerus, forming a hinge joint, which allows flexion and extension. The articulation between the radius and ulna forms a pivot joint, which allows rotational movement. The knee is a complex hinge joint between the femur, tibia, and patella, and its main movements are flexion and extension, although some backward and forward gliding movement and some rotation also occur. The hip is an extremely stable and strong ball-and-socket joint between the femur and the hip bone. The hip's capability of rotation is second only to that of the shoulder joint.

Temporomandibular Joint

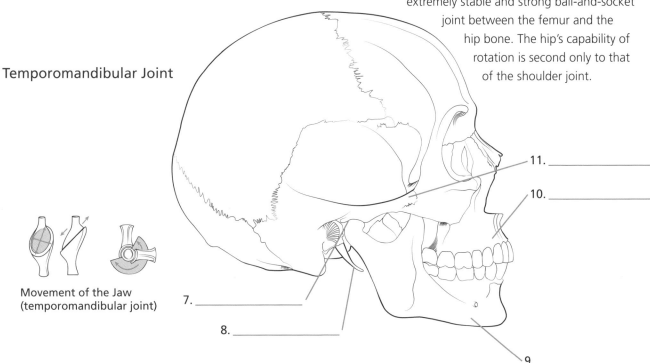

Movement of the Jaw
(temporomandibular joint)

7. _____
8. _____
9. _____
10. _____
11. _____

Answers

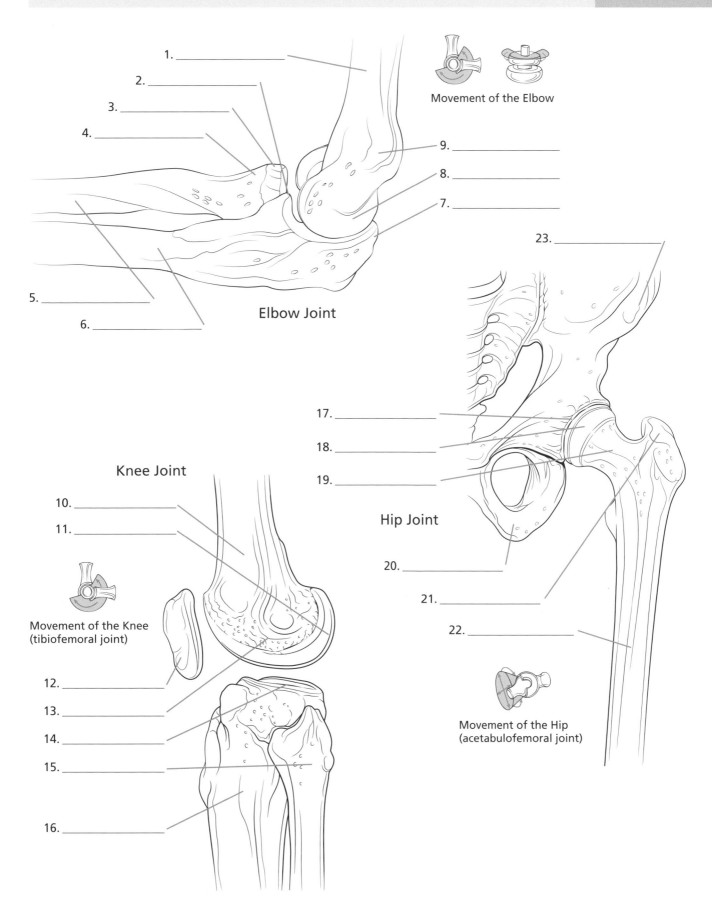

1. _____

2. _____

3. _____

4. _____

5. _____

6. _____

Elbow Joint

Movement of the Elbow

7. _____

8. _____

9. _____

Knee Joint

10. _____

11. _____

Movement of the Knee
(tibiofemoral joint)

12. _____

13. _____

14. _____

15. _____

16. _____

23. _____

17. _____

18. _____

19. _____

Hip Joint

20. _____

21. _____

22. _____

Movement of the Hip
(acetabulofemoral joint)

Answers

1. Humerus, 2. Coronoid process of ulna, 3. Head of radius, 4. Neck of radius, 5. Radius, 6. Ulna, 7. Olecranon, 8. Trochlea of humerus, 9. Medial epicondyle of humerus, 10. Femur, 11. Articular cartilage, 12. Patella, 13. Lateral condyle, 14. Tibial plateau, 15. Fibula, 16. Tibia, 17. Acetabular rim, 18. Head of femur, 19. Neck of femur, 20. Ischium, 21. Greater trochanter, 22. Femur, 23. Illium

Muscular System

Muscular System—
anterior view

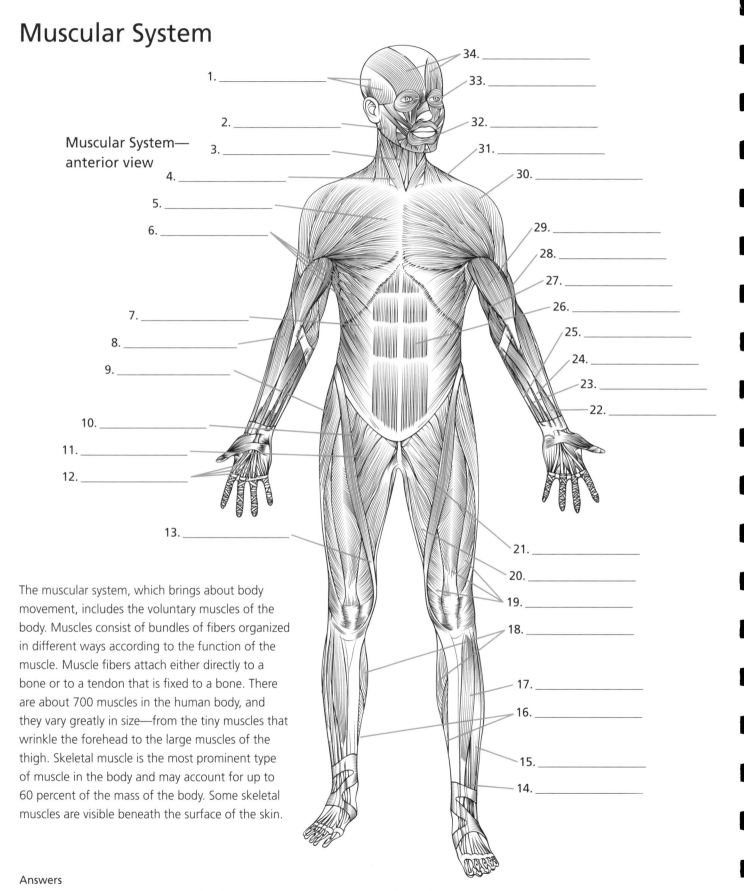

1. _____
2. _____
3. _____
4. _____
5. _____
6. _____
7. _____
8. _____
9. _____
10. _____
11. _____
12. _____
13. _____

34. _____
33. _____
32. _____
31. _____
30. _____
29. _____
28. _____
27. _____
26. _____
25. _____
24. _____
23. _____
22. _____
21. _____
20. _____
19. _____
18. _____
17. _____
16. _____
15. _____
14. _____

The muscular system, which brings about body movement, includes the voluntary muscles of the body. Muscles consist of bundles of fibers organized in different ways according to the function of the muscle. Muscle fibers attach either directly to a bone or to a tendon that is fixed to a bone. There are about 700 muscles in the human body, and they vary greatly in size—from the tiny muscles that wrinkle the forehead to the large muscles of the thigh. Skeletal muscle is the most prominent type of muscle in the body and may account for up to 60 percent of the mass of the body. Some skeletal muscles are visible beneath the surface of the skin.

Answers

1. Temporalis. 2. Masseter. 3. Sternohyoid. 4. Sternocleidomastoid. 5. Pectoralis major. 6. Serratus anterior. 7. External oblique. 8. Brachioradialis. 9. Tensor fascia lata. 10. Iliopsoas. 11. Pectineus. 12. Lumbricals. 13. Sartorius. 14. Extensor hallucis longus. 15. Extensor digitorum longus. 16. Soleus. 17. Tibialis anterior. 18. Gastrocnemius. 19. Quadriceps femoris. 20. Adductor magnus. 21. Adductor longus. 22. Flexor digitorum superficialis. 23. Palmaris longus. 24. Flexor carpi radialis. 25. Flexor carpi ulnaris. 26. Rectus abdominis. 27. Triceps brachii. 28. Brachialis. 29. Biceps brachii. 30. Deltoid. 31. Trapezius. 32. Orbicularis oris. 33. Orbicularis oculi. 34. Frontalis

Muscular System—posterior view

Muscular System—lateral view

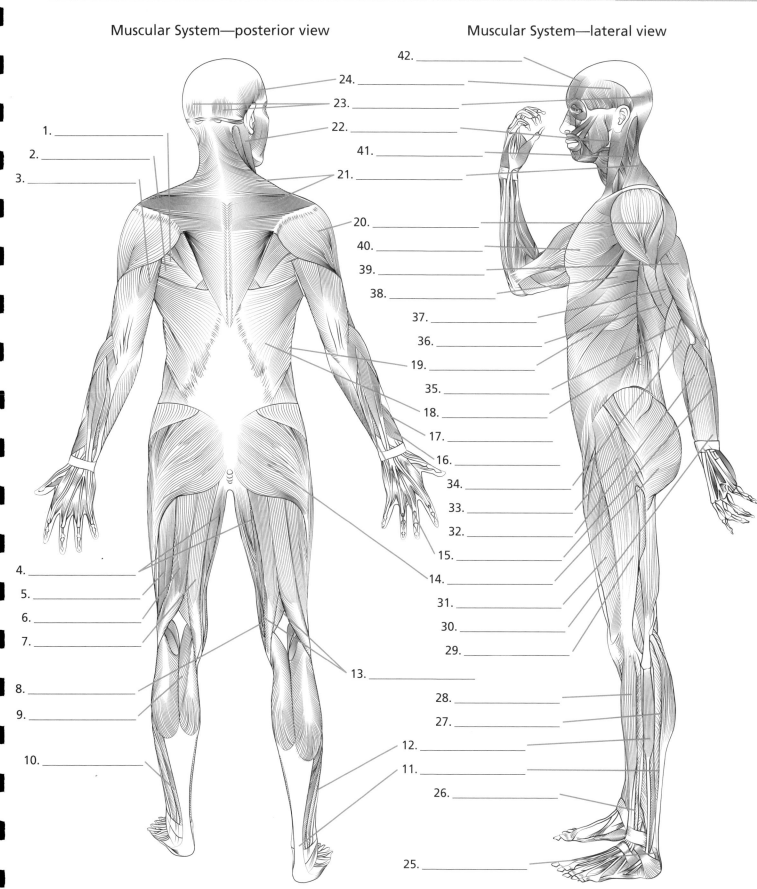

42. _____

24. _____

23. _____

22. _____

41. _____

21. _____

1. _____

2. _____

3. _____

20. _____

40. _____

39. _____

38. _____

37. _____

36. _____

19. _____

35. _____

18. _____

17. _____

16. _____

34. _____

33. _____

32. _____

15. _____

14. _____

4. _____

5. _____

6. _____

7. _____

31. _____

30. _____

29. _____

13. _____

8. _____

9. _____

28. _____

27. _____

12. _____

11. _____

26. _____

10. _____

25. _____

Answers

1. Teres minor, 2. Teres major, 3. Triceps brachii, 4. Adductor magnus, 5. Vastus lateralis, 6. Long head of biceps femoris, 7. Semitendinosus, 8. Gracilis, 9. Gastrocnemius, 10. Soleus, 11. Achilles tendon, 12. Fibularis longus, 13. Semimembranosus, 14. Gluteus maximus, 15. Flexor carpi ulnaris, 16. Extensor pollicis brevis, 17. Abductor pollicis longus, 18. Latissimus dorsi, 19. External oblique, 20. Deltoid, 21. Trapezius, 22. Sternocleidomastoid, 23. Occipitalis, 24. Temporalis, 25. Achilles tendon, 26. Lateral head of gastrocnemius, 27. Tibialis anterior, 28. Tibialis anterior, 29. Iliotibial tract, 30. Extensor carpi ulnaris, 31. Quadriceps (vastus lateralis), 32. Extensor digitorum, 33. Tensor fascia lata, 34. Extensor carpi radialis longus, 35. Brachioradialis, 36. Biceps brachii, 37. Brachialis, 38. Serratus anterior, 39. Lateral head of triceps, 40. Pectoralis major, 41. Levator scapulae, 42. Frontalis

Muscle Types

Muscle fibers are elongated cells containing myofibrils. Each myofibril is made up of thousands of overlapping thin actin filaments and thicker myosin filaments. Where they overlap, actin and myosin filaments are linked by connections (crossbridges) that participate in muscle shortening (contraction). Muscles are classified based on their general shape—some muscles have mainly parallel fibers, and others have oblique fibers. In the case of circular muscles, the fibers are arranged in concentric circles. In common spindle-shaped muscles, all muscle fibers run from one tendon to another. Pennate muscles have a featherlike appearance, and their fibers run obliquely down to the tendon. Some spiral muscles have the capacity to turn half a rotation between their attachments, while others twist around a bone. The shape and arrangement of muscle fibers reflect the function of the muscle.

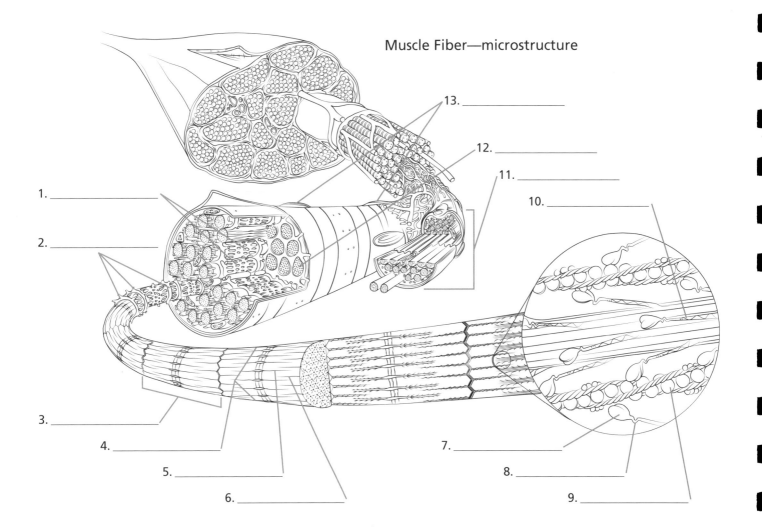

Muscle Fiber—microstructure

1. _____
2. _____
3. _____
4. _____
5. _____
6. _____
7. _____
8. _____
9. _____
10. _____
11. _____
12. _____
13. _____

Muscle Shapes

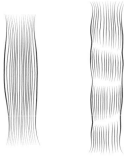

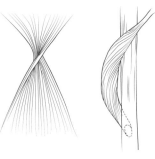

1. _____

2. _____

3. _____

4. _____

5. _____

6. _____

7. _____

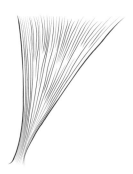

8. _____

9. _____

10. _____

11. _____

12. _____

13. _____

14. _____

15. _____

16. _____

Muscles of the Head and Neck

Superficial and Deep Muscles of the Head and Neck— anterior view

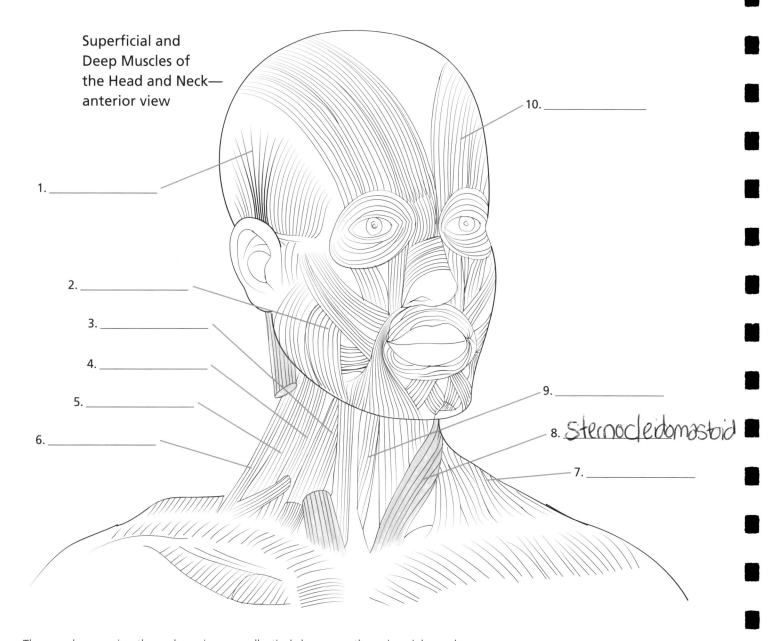

1. _____
2. _____
3. _____
4. _____
5. _____
6. _____
7. _____
8. Sternocleidomastoid
9. _____
10. _____

The muscles covering the scalp region are collectively known as the epicranial muscles and include the occipitalis, frontalis, and temporoparietal muscles. The neck muscles attach to the front, back, and sides of the vertebrae, producing forward, backward, and sideways movements. Those with an oblique orientation also produce rotation (turning). The largest musculature lies to the back. Some of these muscles are exclusively related to moving the head and neck (for example, splenius capitis and cervicis and semispinalis capitis and cervicis), while others are related to moving the shoulder (for example, trapezius, levator scapulae) or raising the upper two ribs (scalene muscles).

Answers

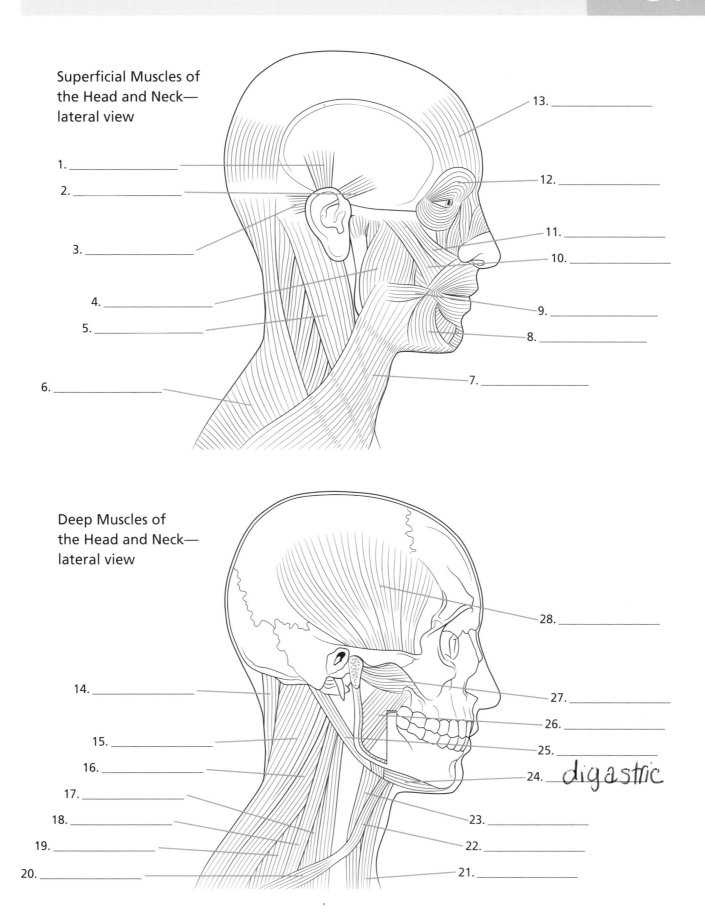

Superficial Muscles of the Head and Neck— lateral view

1. _____
2. _____
3. _____
4. _____
5. _____
6. _____

13. _____
12. _____
11. _____
10. _____
9. _____
8. _____
7. _____

Deep Muscles of the Head and Neck— lateral view

14. _____
15. _____
16. _____
17. _____
18. _____
19. _____
20. _____

28. _____
27. _____
26. _____
25. _____
24. *digastric*
23. _____
22. _____
21. _____

Answers

1. Auricularis superior, 2. Auricularis anterior, 3. Auricularis posterior, 4. Masseter, 5. Sternocleidomastoid, 6. Trapezius, 7. Platysma, 8. Depressor anguli oris, 9. Risorius, 10. Zygomaticus major, 11. Zygomaticus minor, 12. Orbicularis oculi, 13. Frontalis, 14. Semispinalis capitis, 15. Splenius capitis, 16. Levator scapulae, 17. Scalenus anterior, 18. Scalenus medius, 19. Scalenus posterior, 20. Omohyoid (inferior belly), 21. Sternohyoid, 22. Omohyoid (superior belly), 23. Thyrohyoid, 24. Digastric (anterior belly), 25. Digastric (posterior belly), 26. Medial pterygoid, 27. Lateral pterygoid, 28. Temporalis

Muscles of the Head and Neck

Superficial Muscles of the Jaw— lateral view

The superficial muscles of the jaw (the temporalis and the masseter) in combination with the deep muscles (the medial pterygoid and the lateral pterygoid) are the muscles of mastication. Combined, they allow the jaw to move up, down, to the side, forward, and backward. The temporalis muscle arises at the temporal fossa and temporal fascia and inserts at the coronoid process of the mandible; the masseter arises at the inferior border of the zygomatic arch and inserts at the ramus and coronoid process of the mandible. The medial pterygoid (deep part) and lateral pterygoid (inferior head) arise at the lateral pterygoid plate; the medial pterygoid (superficial part) at the tuberosity of the maxilla; and the lateral pterygoid (superior head) at the infratemporal surface of the greater wing of the sphenoid bone and the pterygoid ridge. The medial pterygoid inserts at the medial surface of the mandibular ramus and the angle of the mandible; and the lateral pterygoid inserts at the temporomandibular joint and the neck of the mandible.

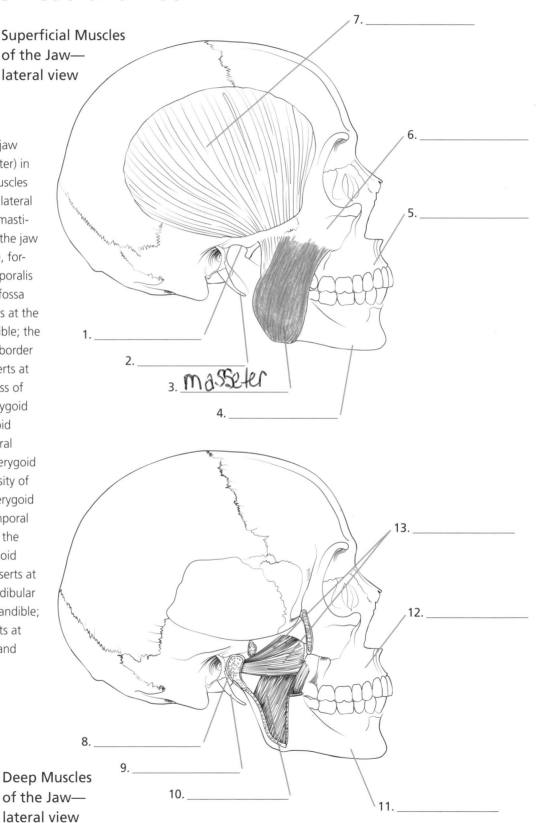

7. _____

6. _____

5. _____

1. _____

2. _____

3. _masseter_

4. _____

13. _____

12. _____

8. _____

9. _____

10. _____

11. _____

Deep Muscles of the Jaw— lateral view

Muscles of Facial Expression— lateral view

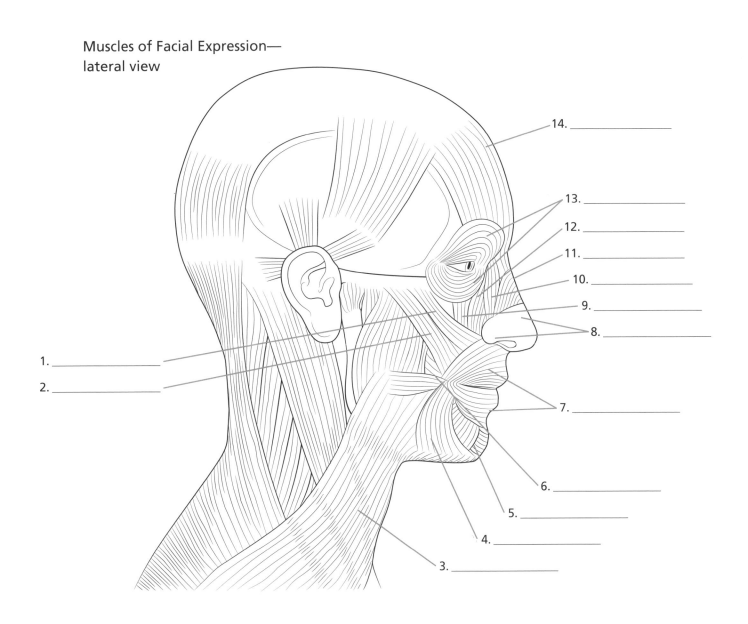

14. _____

13. _____

12. _____

11. _____

10. _____

9. _____

8. _____

1. _____

2. _____

7. _____

6. _____

5. _____

4. _____

3. _____

The muscles of the face are responsible for facial expression. There is a circular muscle around the mouth, and one around each eye. Other muscles spread out over the face from the edge of the circular muscles. The muscles of the eye region are the frontalis, orbicularis oculi, and procerus. The muscles found in the nasal region are the procerus, levator labii superioris alaeque nasi, and nasalis. The muscles of the mouth are the zygomaticus major, zygomaticus minor, levator labii superioris, levator labii superioris alaeque nasi, orbicularis oris, buccinator, depressor labii inferioris, and depressor anguli oris.

Muscles of the Head and Neck

The movement of the eyeball is controlled by six muscles (superior rectus, medial rectus, lateral rectus, inferior rectus, superior oblique, and inferior oblique). These muscles cause the eye to look up, down, left, and right. The levator palpebrae superioris is involved in movement of the upper eyelid. The trochlea is a hook of ligament that acts as a fulcrum for the superior oblique muscle. The inferior rectus, medial rectus, superior rectus, lateral rectus, and levator palpebrae superioris muscles all arise at the dural sheath around the optic nerve; the superior oblique arises at the sphenoid bone; and the inferior oblique at the maxillary bone. The six ocular muscles insert on the sclera, and the levator palpebrae superioris at the tarsal plate of the upper eyelid.

Muscles of the Eye— superior view

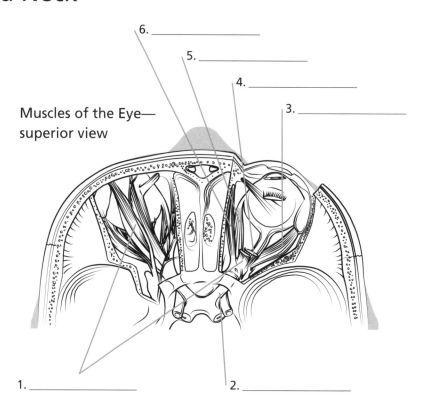

6. _____

5. _____

4. _____

3. _____

1. _____

2. _____

Muscles of the Eye— lateral view

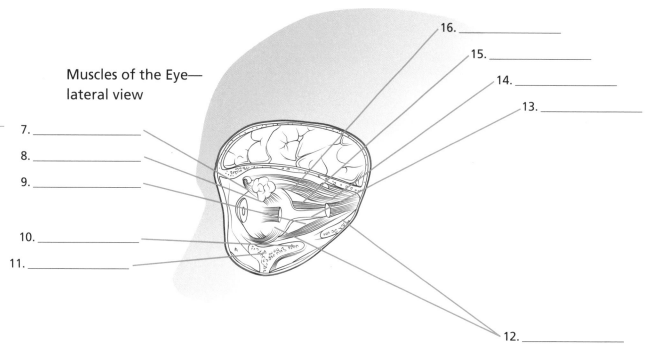

7. _____

8. _____

9. _____

10. _____

11. _____

16. _____

15. _____

14. _____

13. _____

12. _____

Answers

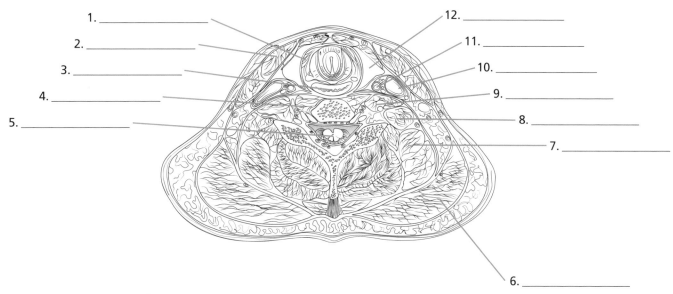

1. _____
2. _____
3. _____
4. _____
5. _____

12. _____
11. _____
10. _____
9. _____
8. _____
7. _____
6. _____

Muscles of the Neck—transverse section

The transverse section of the neck shows the relationship among the components of the neck column. The neck can be divided into two major columns. At the back, the nuchal region includes the cervical vertebrae and their supporting musculature. In front, the neck includes a visceral column containing the larynx, trachea, pharynx, and esophagus. The constrictor muscles of the pharynx are involved in the digestive process, being responsible for moving food down to the esophagus. The stylopharyngeus and the deeper muscles, the palatopharyngeus and the salpingopharyngeus, are involved in elevating the larynx. The superior constrictor muscle arises at the pterygoid plate of the sphenoid bone; the middle constrictor at the horns of the hyoid bone; and the inferior constrictor at the cricoid and thyroid cartilages of the larynx. All three constrictor muscles insert at the median raphe of the pharynx. The stylopharyngeus arises at the styloid process; the salpingopharyngeus at the cartilage of the auditory tube; and the palatopharyngeus at the hard palate in the oral cavity. These three muscles insert at the thyroid cartilage of the larynx.

Muscles of the Pharynx— posterior view

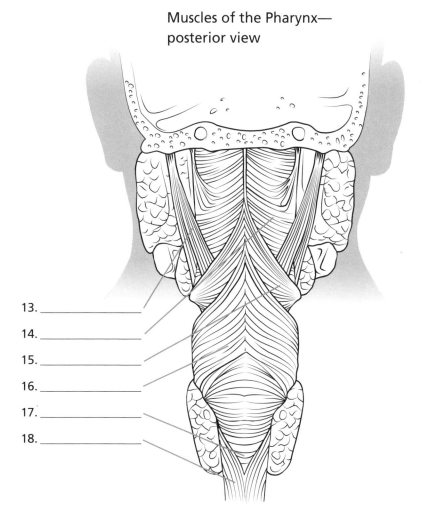

13. _____
14. _____
15. _____
16. _____
17. _____
18. _____

Answers

Muscles of the Back

Superficial Muscles of the Back— posterior view

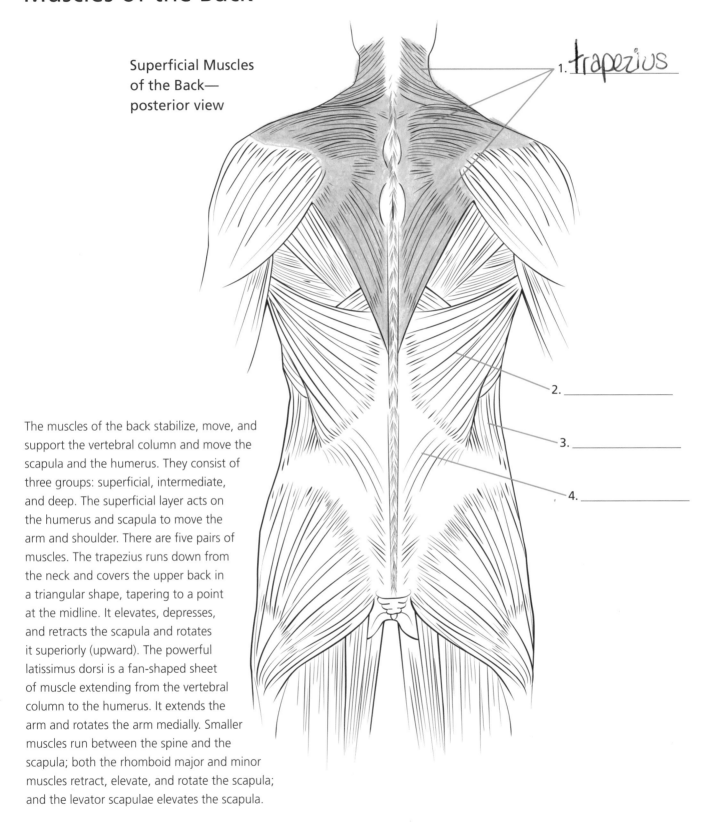

1. *trapezius*

2. _____

3. _____

4. _____

The muscles of the back stabilize, move, and support the vertebral column and move the scapula and the humerus. They consist of three groups: superficial, intermediate, and deep. The superficial layer acts on the humerus and scapula to move the arm and shoulder. There are five pairs of muscles. The trapezius runs down from the neck and covers the upper back in a triangular shape, tapering to a point at the midline. It elevates, depresses, and retracts the scapula and rotates it superiorly (upward). The powerful latissimus dorsi is a fan-shaped sheet of muscle extending from the vertebral column to the humerus. It extends the arm and rotates the arm medially. Smaller muscles run between the spine and the scapula; both the rhomboid major and minor muscles retract, elevate, and rotate the scapula; and the levator scapulae elevates the scapula.

Answers

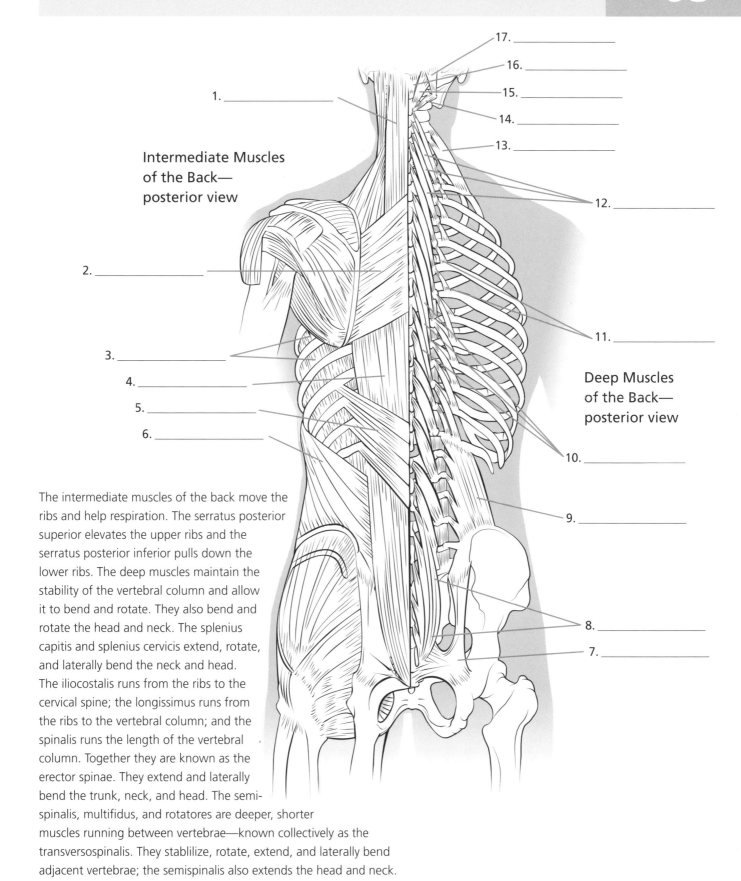

Intermediate Muscles of the Back— posterior view

1. _____
2. _____
3. _____
4. _____
5. _____
6. _____

17. _____
16. _____
15. _____
14. _____
13. _____
12. _____
11. _____

Deep Muscles of the Back— posterior view

10. _____
9. _____
8. _____
7. _____

The intermediate muscles of the back move the ribs and help respiration. The serratus posterior superior elevates the upper ribs and the serratus posterior inferior pulls down the lower ribs. The deep muscles maintain the stability of the vertebral column and allow it to bend and rotate. They also bend and rotate the head and neck. The splenius capitis and splenius cervicis extend, rotate, and laterally bend the neck and head. The iliocostalis runs from the ribs to the cervical spine; the longissimus runs from the ribs to the vertebral column; and the spinalis runs the length of the vertebral column. Together they are known as the erector spinae. They extend and laterally bend the trunk, neck, and head. The semi- spinalis, multifidus, and rotatores are deeper, shorter muscles running between vertebrae—known collectively as the transversospinalis. They stablilize, rotate, extend, and laterally bend adjacent vertebrae; the semispinalis also extends the head and neck.

Answers

1. Semispinalis capitis, 2. Rhomboid major, 3. External intercostal, 4. Erector spinae, 5. Serratus posterior inferior, 6. Internal oblique, 7. Sacrotuberous ligament, 8. Multifidus, 9. Quadratus lumborum, 10. Semispinalis thoracis, 11. Levatores costarum, 12. Semispinalis cervicis, 13. Scalenus posterior, 14. Oblique capitis inferior, 15. Rectus capitis posterior major, 16. Rectus capitis posterior minor, 17. Oblique capitis superior.

Muscles of the Thorax and Abdomen

Superficial and Deep Muscles of the Thorax and Abdomen— anterior view

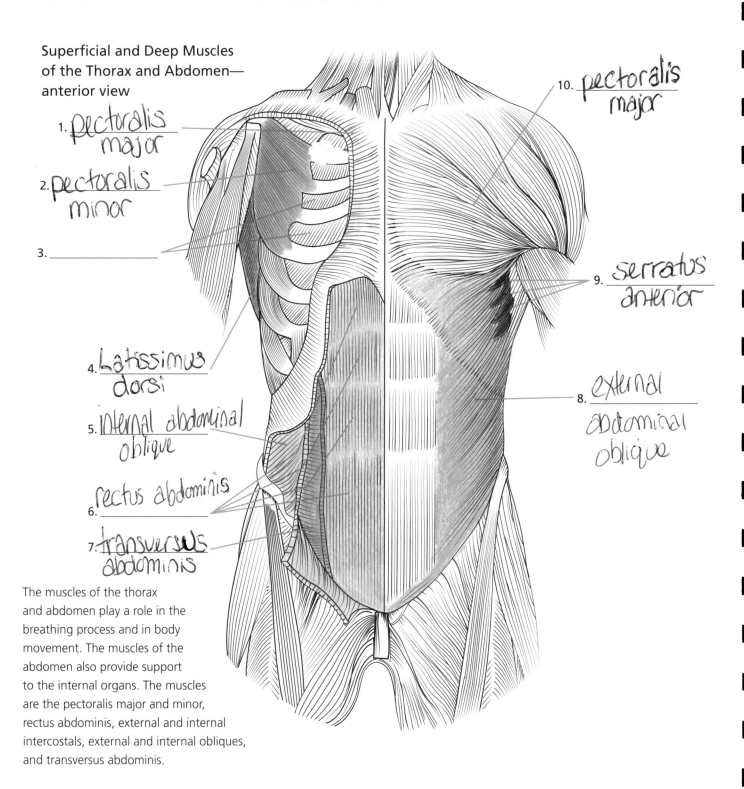

1. pectoralis major
2. pectoralis minor
3. _____
4. Latissimus dorsi
5. internal abdominal oblique
6. rectus abdominis
7. transversus abdominis

8. external abdominal oblique
9. serratus anterior
10. pectoralis major

The muscles of the thorax and abdomen play a role in the breathing process and in body movement. The muscles of the abdomen also provide support to the internal organs. The muscles are the pectoralis major and minor, rectus abdominis, external and internal intercostals, external and internal obliques, and transversus abdominis.

Answers

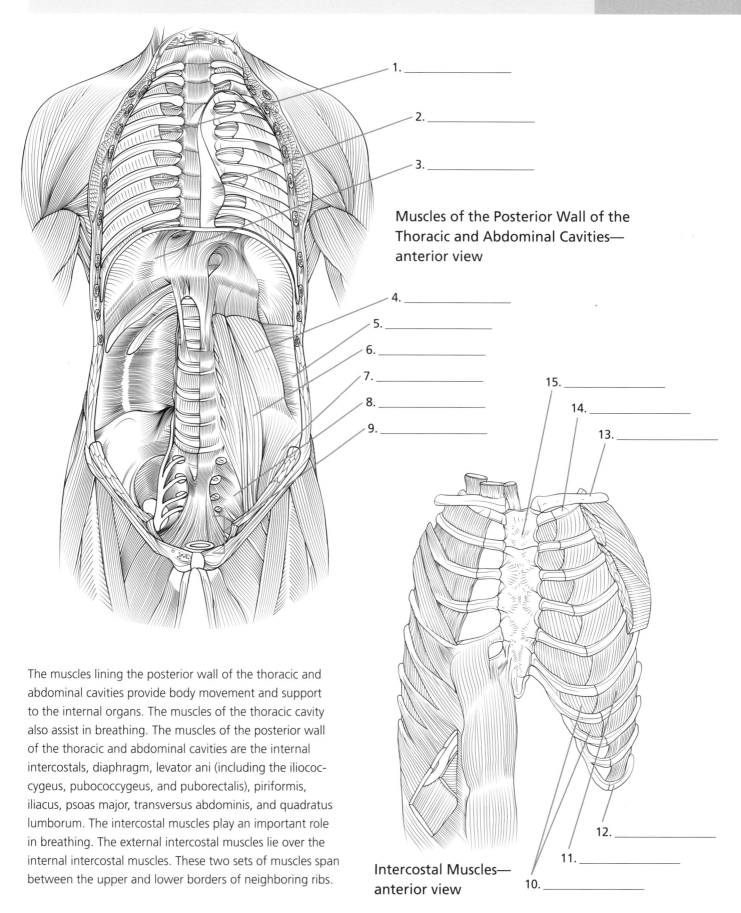

Muscles of the Posterior Wall of the Thoracic and Abdominal Cavities— anterior view

1. _____

2. _____

3. _____

4. _____

5. _____

6. _____

7. _____

8. _____

9. _____

15. _____

14. _____

13. _____

12. _____

11. _____

10. _____

Intercostal Muscles— anterior view

The muscles lining the posterior wall of the thoracic and abdominal cavities provide body movement and support to the internal organs. The muscles of the thoracic cavity also assist in breathing. The muscles of the posterior wall of the thoracic and abdominal cavities are the internal intercostals, diaphragm, levator ani (including the iliococcygeus, pubococcygeus, and puborectalis), piriformis, iliacus, psoas major, transversus abdominis, and quadratus lumborum. The intercostal muscles play an important role in breathing. The external intercostal muscles lie over the internal intercostal muscles. These two sets of muscles span between the upper and lower borders of neighboring ribs.

Answers

1. Internal intercostal and innermost intercostal. 2. Mediastinal pleura. 3. Diaphragm. 4. Quadratus lumborum. 5. Transversus abdominis. 6. Psoas major. 7. Iliacus. 8. Piriformis, 9. Levator ani (iliococcygeus, pubococcygeus, and puborectalis). 10. Internal intercostals. 11. External intercostal. 12. Tenth rib. 13. Clavicle. 14. First rib. 15. Manubrium of sternum

Muscles of the Thorax and Abdomen

The diaphragm is a sheet of muscle that separates the chest cavity from the abdominal cavity, and is pierced by several structures that pass between the two cavities. The three largest of these structures are the esophagus, the aorta, and the inferior vena cava. The heart and lungs lie on the upper convex surface of the diaphragm, with the pericardial sac firmly attached to the upper surface of the central tendon of the diaphragm, which is the central part of the diaphragm.

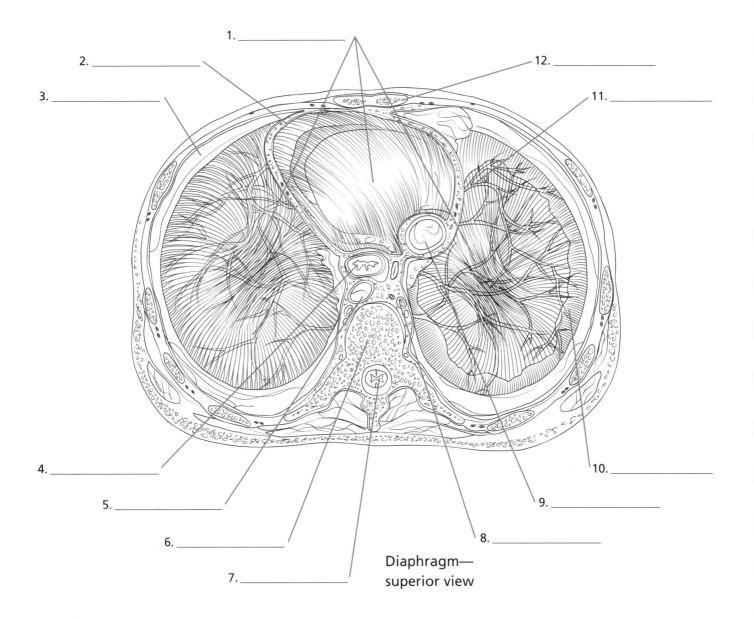

1. _____

2. _____

3. _____

12. _____

11. _____

4. _____

5. _____

6. _____

7. _____

8. _____

9. _____

10. _____

Diaphragm—
superior view

The inferior surface of the diaphragm forms the roof of the abdominal cavity and lies over the stomach, on the left, and the liver, on the right. The diaphragm arises at the vertebral column, rib pairs 7–12 along the side of the chest, and the xiphoid process of the sternum and inserts at the central tendon of the diaphragm.

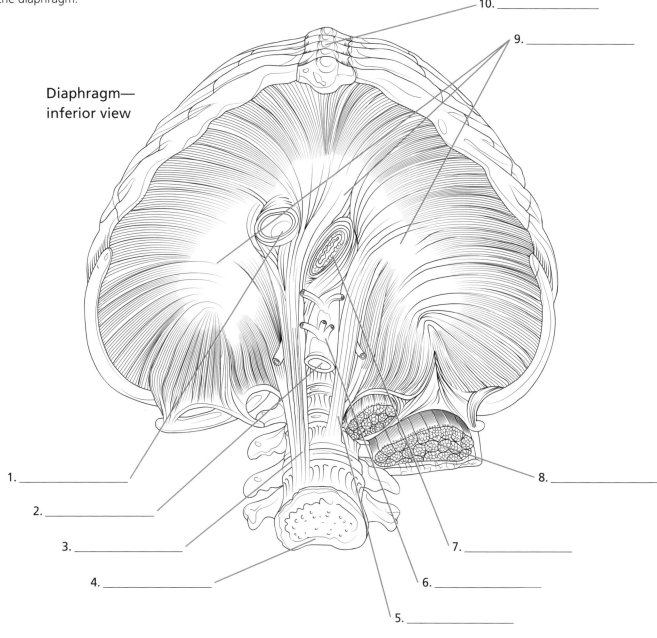

Diaphragm—
inferior view

10. _____

9. _____

1. _____

2. _____

3. _____

4. _____

5. _____

6. _____

7. _____

8. _____

Answers

Pelvic Floor Muscles and Muscles of the Perineum

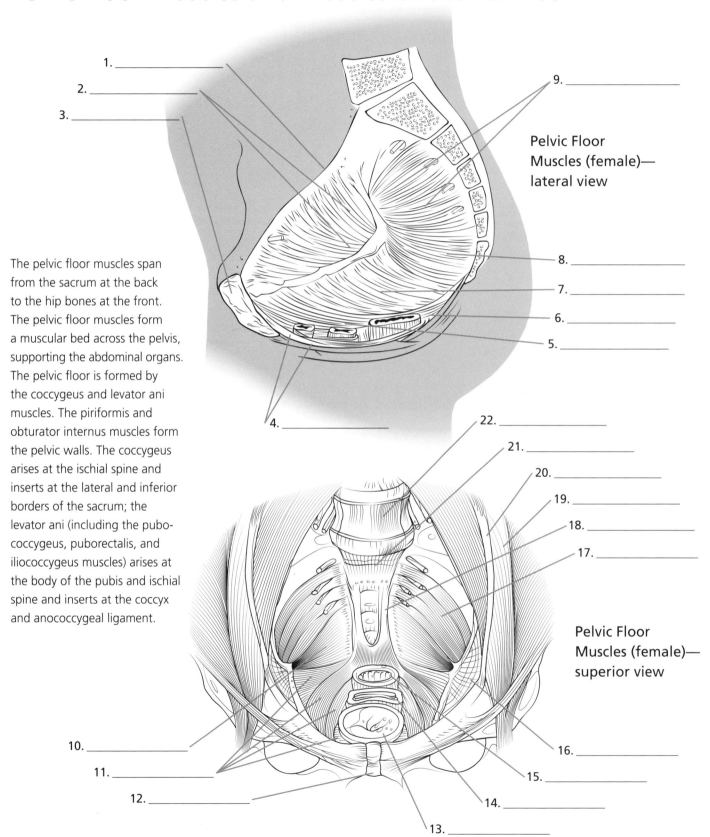

1. _____

2. _____

3. _____

4. _____

Pelvic Floor Muscles (female)— lateral view

5. _____

6. _____

7. _____

8. _____

9. _____

The pelvic floor muscles span from the sacrum at the back to the hip bones at the front. The pelvic floor muscles form a muscular bed across the pelvis, supporting the abdominal organs. The pelvic floor is formed by the coccygeus and levator ani muscles. The piriformis and obturator internus muscles form the pelvic walls. The coccygeus arises at the ischial spine and inserts at the lateral and inferior borders of the sacrum; the levator ani (including the pubococcygeus, puborectalis, and iliococcygeus muscles) arises at the body of the pubis and ischial spine and inserts at the coccyx and anococcygeal ligament.

Pelvic Floor Muscles (female)— superior view

10. _____

11. _____

12. _____

13. _____

14. _____

15. _____

16. _____

17. _____

18. _____

19. _____

20. _____

21. _____

22. _____

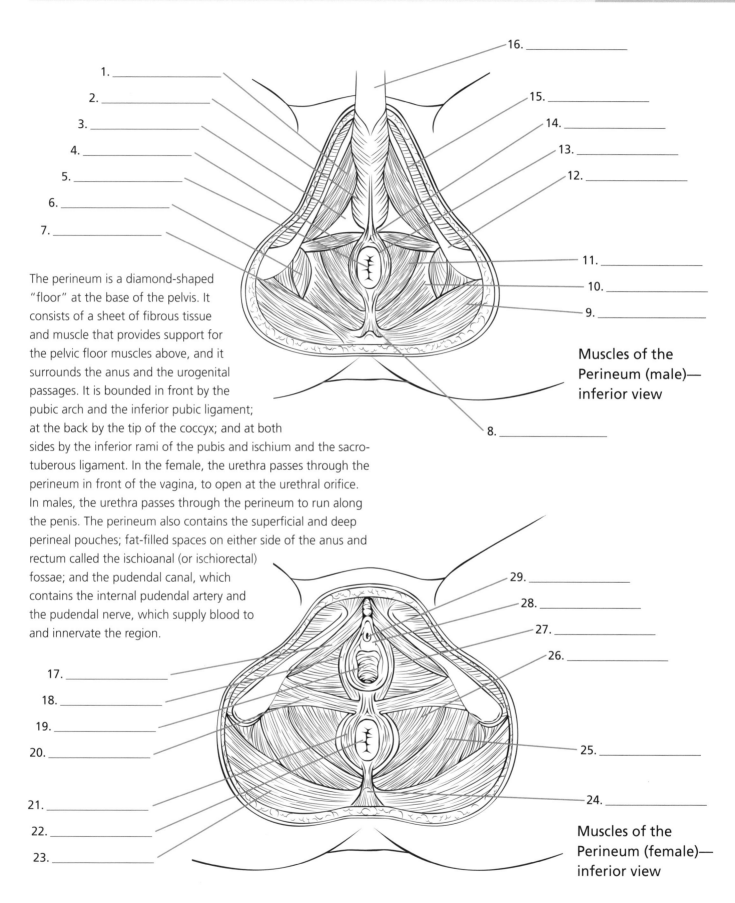

The perineum is a diamond-shaped "floor" at the base of the pelvis. It consists of a sheet of fibrous tissue and muscle that provides support for the pelvic floor muscles above, and it surrounds the anus and the urogenital passages. It is bounded in front by the pubic arch and the inferior pubic ligament; at the back by the tip of the coccyx; and at both sides by the inferior rami of the pubis and ischium and the sacro-tuberous ligament. In the female, the urethra passes through the perineum in front of the vagina, to open at the urethral orifice. In males, the urethra passes through the perineum to run along the penis. The perineum also contains the superficial and deep perineal pouches; fat-filled spaces on either side of the anus and rectum called the ischioanal (or ischiorectal) fossae; and the pudendal canal, which contains the internal pudendal artery and the pudendal nerve, which supply blood to and innervate the region.

1. _____
2. _____
3. _____
4. _____
5. _____
6. _____
7. _____

16. _____
15. _____
14. _____
13. _____
12. _____
11. _____
10. _____
9. _____
8. _____

Muscles of the Perineum (male)—inferior view

17. _____
18. _____
19. _____
20. _____
21. _____
22. _____
23. _____

29. _____
28. _____
27. _____
26. _____
25. _____
24. _____

Muscles of the Perineum (female)—inferior view

Answers

Muscles of the Shoulder

Superficial and Deep
Muscles of the Shoulder—
posterior view

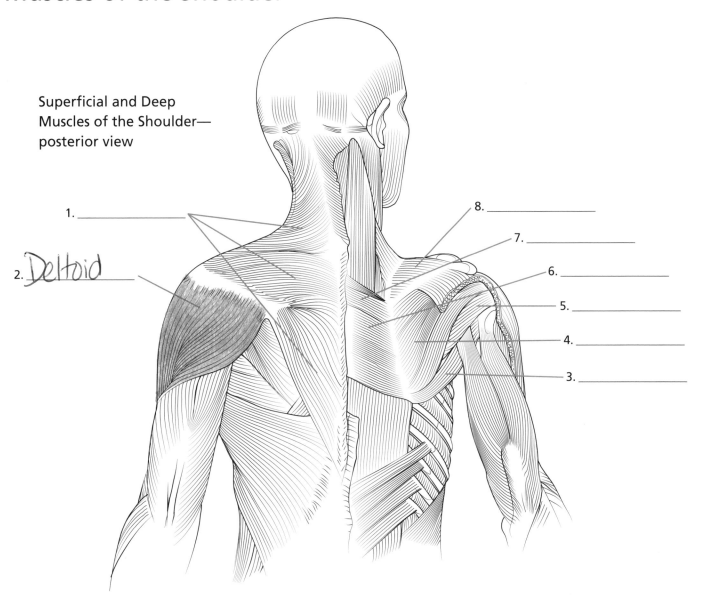

1. _____

2. Deltoid _____

8. _____

7. _____

6. _____

5. _____

4. _____

3. _____

The pectoral girdle has two groups of muscles: one group attaches the humerus to the shoulder girdle, and the other attaches the shoulder girdle to the trunk. The trapezius, deltoid, and latissimus dorsi are superficial muscles; the infraspinatus, rhomboid major and minor, and supraspinatus are deep muscles. The trapezius arises at the occipital bone and spinous processes from the lowest cervical vertebra to the lowest thoracic vertebra (C7–T12) and inserts at the clavicle and acromion and spine of the scapula; the deltoid arises at the clavicle and acromion and spine of the scapula and inserts at the deltoid tuberosity of the humerus; the latissimus dorsi arises at the spinous processes of the lower vertebrae, thoracolumbar fascia, and rib pairs 8–12 and inserts at the intertubercular groove of the humerus; and the infraspinatus and supraspinatus arise at the scapula and insert at the greater tubercle of the humerus. The rhomboid major arises at the spinous processes of thoracic vertebrae T2–T5; the rhomboid minor arises at the spinous processes of the last cervical vertebra and first thoracic vertebra (C7–T1); and both insert at the medial border of the scapula.

Answers

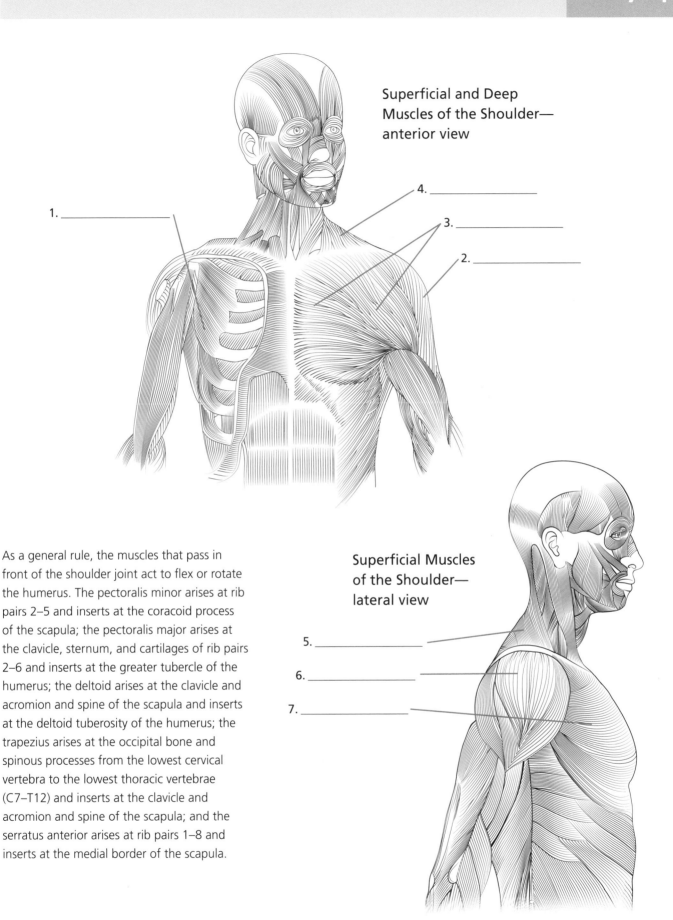

Superficial and Deep Muscles of the Shoulder— anterior view

1. _____

4. _____

3. _____

2. _____

As a general rule, the muscles that pass in front of the shoulder joint act to flex or rotate the humerus. The pectoralis minor arises at rib pairs 2–5 and inserts at the coracoid process of the scapula; the pectoralis major arises at the clavicle, sternum, and cartilages of rib pairs 2–6 and inserts at the greater tubercle of the humerus; the deltoid arises at the clavicle and acromion and spine of the scapula and inserts at the deltoid tuberosity of the humerus; the trapezius arises at the occipital bone and spinous processes from the lowest cervical vertebra to the lowest thoracic vertebrae (C7–T12) and inserts at the clavicle and acromion and spine of the scapula; and the serratus anterior arises at rib pairs 1–8 and inserts at the medial border of the scapula.

Superficial Muscles of the Shoulder— lateral view

5. _____

6. _____

7. _____

Answers

1. Pectoralis minor, 2. Deltoid, 3. Pectoralis major, 4. Trapezius, 5. Trapezius, 6. Deltoid, 7. Pectoralis major

Muscles of the Upper Limb

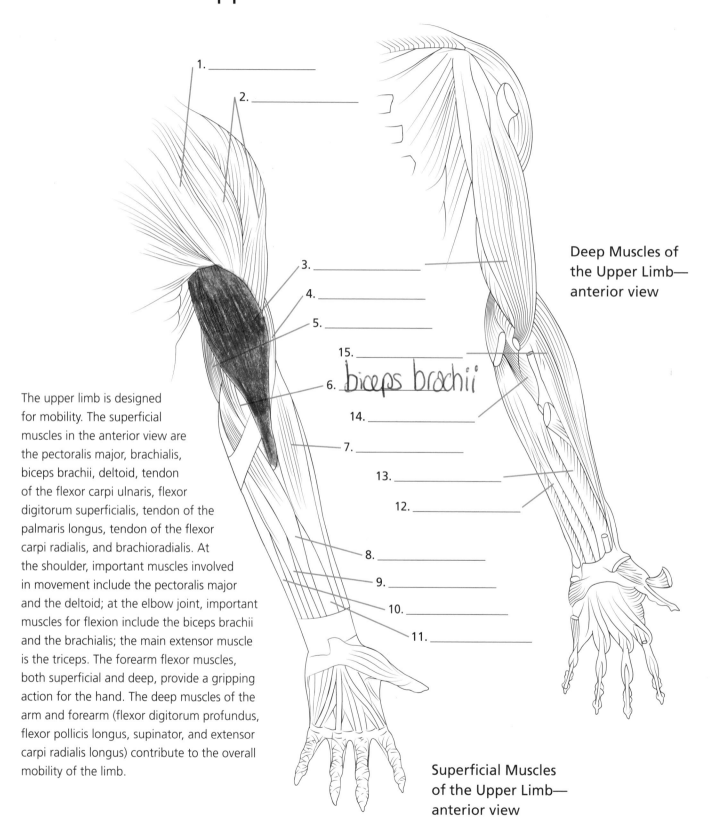

1. _____

2. _____

Deep Muscles of the Upper Limb— anterior view

3. _____

4. _____

5. _____

15. _____

6. *biceps brachii*

14. _____

The upper limb is designed for mobility. The superficial muscles in the anterior view are the pectoralis major, brachialis, biceps brachii, deltoid, tendon of the flexor carpi ulnaris, flexor digitorum superficialis, tendon of the palmaris longus, tendon of the flexor carpi radialis, and brachioradialis. At the shoulder, important muscles involved in movement include the pectoralis major and the deltoid; at the elbow joint, important muscles for flexion include the biceps brachii and the brachialis; the main extensor muscle is the triceps. The forearm flexor muscles, both superficial and deep, provide a gripping action for the hand. The deep muscles of the arm and forearm (flexor digitorum profundus, flexor pollicis longus, supinator, and extensor carpi radialis longus) contribute to the overall mobility of the limb.

7. _____

13. _____

12. _____

8. _____

9. _____

10. _____

11. _____

Superficial Muscles of the Upper Limb— anterior view

Answers

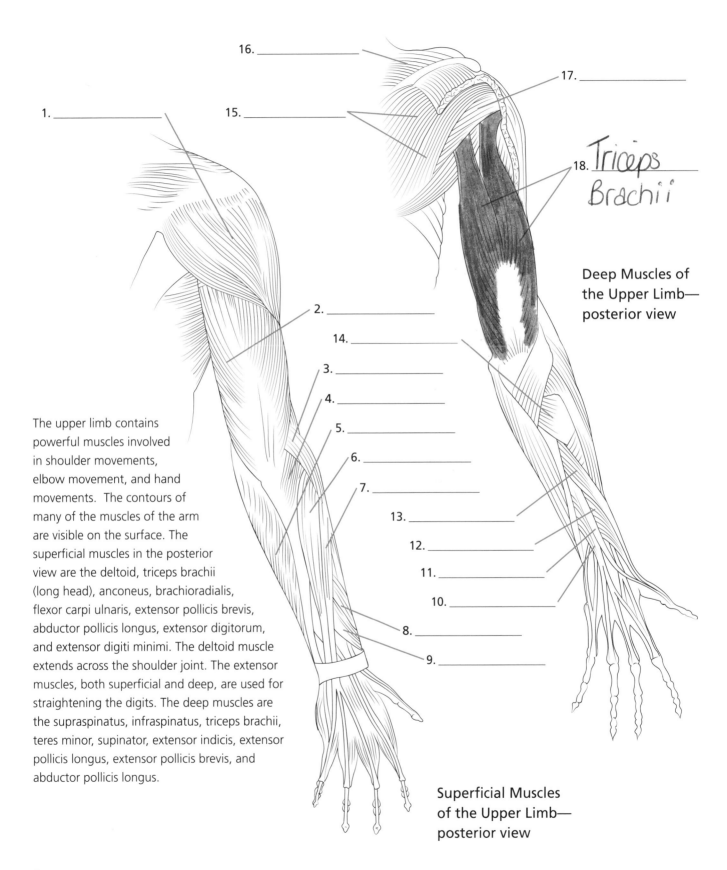

16. _____

17. _____

1. _____

15. _____

18. Triceps Brachii

Deep Muscles of the Upper Limb— posterior view

2. _____

14. _____

3. _____

4. _____

5. _____

6. _____

7. _____

13. _____

12. _____

11. _____

10. _____

8. _____

9. _____

The upper limb contains powerful muscles involved in shoulder movements, elbow movement, and hand movements. The contours of many of the muscles of the arm are visible on the surface. The superficial muscles in the posterior view are the deltoid, triceps brachii (long head), anconeus, brachioradialis, flexor carpi ulnaris, extensor pollicis brevis, abductor pollicis longus, extensor digitorum, and extensor digiti minimi. The deltoid muscle extends across the shoulder joint. The extensor muscles, both superficial and deep, are used for straightening the digits. The deep muscles are the supraspinatus, infraspinatus, triceps brachii, teres minor, supinator, extensor indicis, extensor pollicis longus, extensor pollicis brevis, and abductor pollicis longus.

Superficial Muscles of the Upper Limb— posterior view

Answers

1. Deltoid, 2. Triceps brachii (long head), 3. Brachioradialis, 4. Anconeus, 5. Flexor carpi ulnaris, 6. Extensor digiti minimi, 7. Extensor digitorum, 8. Abductor pollicis longus, 9. Extensor pollicis brevis, 10. Extensor indicis, 11. Extensor pollicis longus, 12. Extensor pollicis brevis, 13. Abductor pollicis longus, 14. Supinator, 15. Infraspinatus, 16. Supraspinatus, 17. Teres minor, 18. Triceps brachii

Muscles of the Upper Limb

The lateral view of the superficial muscles of the upper limb shows the deltoid, lateral head of the triceps, brachialis, biceps brachii, brachioradialis, extensor carpi radialis longus, extensor digitorum, flexor carpi ulnaris, and extensor carpi ulnaris. The lateral view of the elbow shows the muscles of the arm and the extensor muscles of the forearm. The triceps brachii causes elbow extension, while the biceps brachii is responsible for elbow flexion and participates in supination of the forearm. The brachialis and brachioradialis also contribute to elbow flexion.

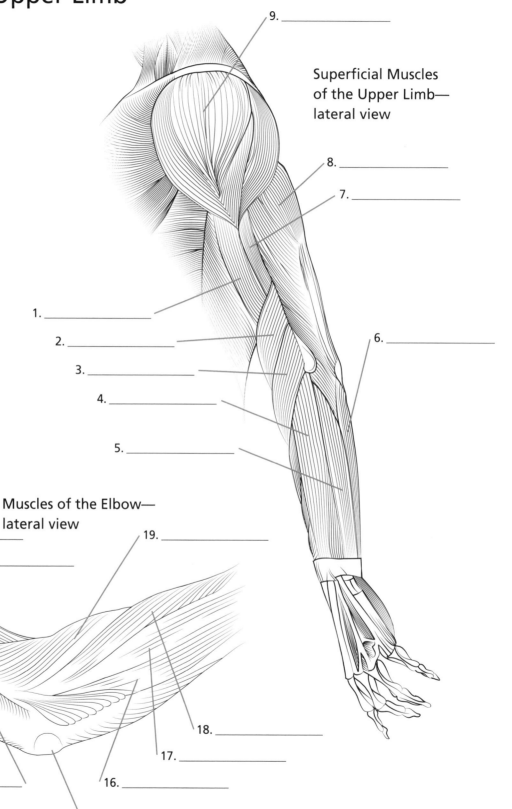

9. _____

Superficial Muscles of the Upper Limb— lateral view

8. _____

7. _____

1. _____

2. _____

3. _____

4. _____

5. _____

6. _____

Muscles of the Elbow— lateral view

12. _____

11. _____

10. _____

19. _____

18. _____

17. _____

16. _____

15. _____

14. _____

13. _____

Answers

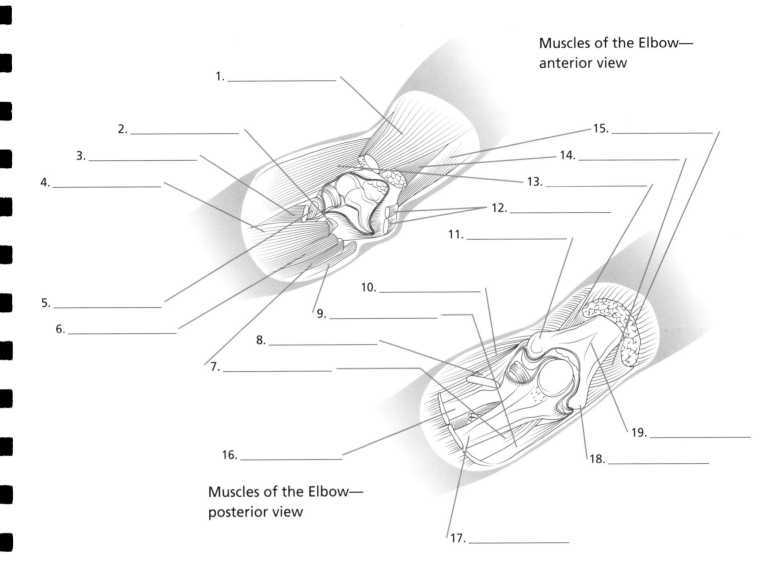

Muscles of the Elbow—anterior view

1. _____

2. _____

3. _____

4. _____

5. _____

6. _____

7. _____

8. _____

9. _____

10. _____

11. _____

12. _____

13. _____

14. _____

15. _____

16. _____

17. _____

18. _____

19. _____

Muscles of the Elbow—posterior view

The anterior view of the muscles of the elbow shows the triceps brachii, biceps brachii, brachioradialis, pronator teres, flexor carpi radialis, biceps brachii tendon, brachialis, palmaris longus, flexor digitorum superficialis, flexor carpi ulnaris, brachialis tendon, and common flexor tendon. The posterior view shows the brachialis, triceps brachii, brachioradialis, extensor carpi radialis brevis, common extensor tendon, flexor digitorum superficialis, and flexor carpi ulnaris. The elbow joint separates the humerus of the upper arm from the forearm bones—the radius and ulna. This relatively stable synovial hinge joint allows the forearm to flex and extend at the elbow. Important muscles for flexion include the biceps brachii and brachialis, while the main extensor muscle is the triceps brachii. The forearm muscles, extending from the elbow to the hand, allow rotation of the forearm, from supination to pronation, flexion of the hand at the wrist, and flexion of the digits. Located at the back of the forearm are the muscles dedicated to extension of the hand at the wrist and flexion of the digits.

Answers

1. Biceps brachii, 2. Brachialis tendon, 3. Pronator teres, 4. Flexor carpi radialis, 5. Biceps brachii tendon, 6. Palmaris longus, 7. Flexor digitorum superficialis, 8. Common extensor tendon, 9. Flexor carpi ulnaris, 10. Extensor carpi radialis brevis, 11. Lateral epicondyle of humerus, 12. Common flexor tendons, 13. Brachioradialis, 14. Brachialis, 15. Triceps brachii, 16. Radius, 17. Ulna, 18. Medial epicondyle, 19. Humerus

Muscles of the Upper Limb

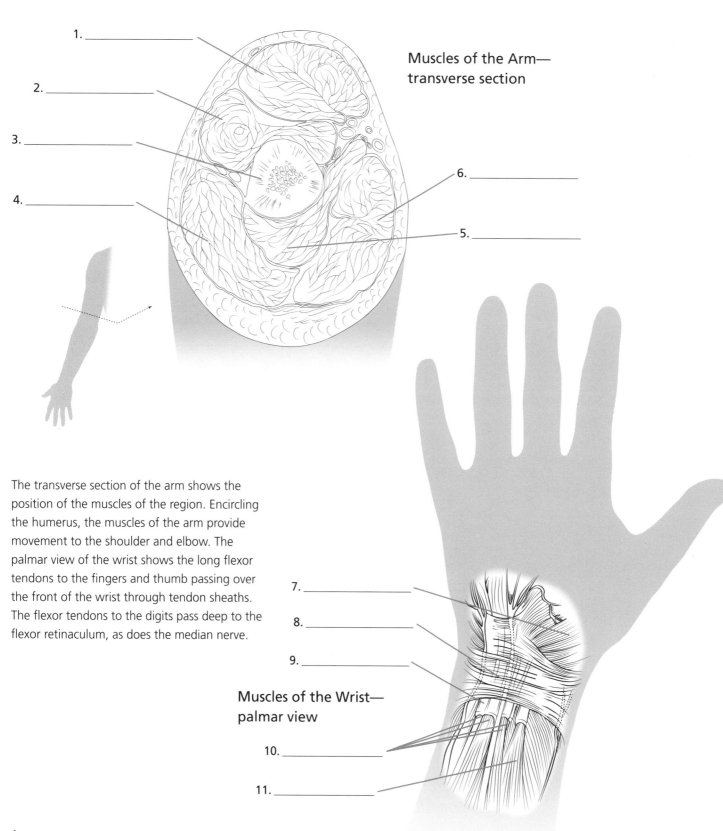

1. _____

2. _____

3. _____

4. _____

Muscles of the Arm—
transverse section

6. _____

5. _____

The transverse section of the arm shows the position of the muscles of the region. Encircling the humerus, the muscles of the arm provide movement to the shoulder and elbow. The palmar view of the wrist shows the long flexor tendons to the fingers and thumb passing over the front of the wrist through tendon sheaths. The flexor tendons to the digits pass deep to the flexor retinaculum, as does the median nerve.

7. _____

8. _____

9. _____

Muscles of the Wrist—
palmar view

10. _____

11. _____

Answers

1. Biceps brachii, 2. Brachialis, 3. Humerus, 4. Triceps brachii (lateral head), 5. Triceps brachii (medial head), 6. Triceps brachii (long head), 7. Thenar muscles, 8. Flexor retinaculum, 9. Tendinous sheath of flexor digitorum superficialis, 10. Flexor digitorum superficialis tendons, 11. Flexor carpi radialis

Muscles of the Wrist and Hand— dorsal view

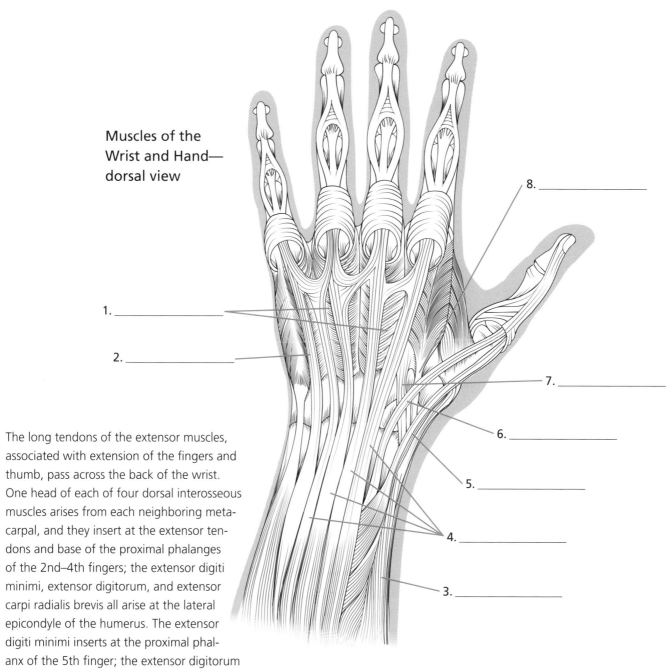

8. _____

1. _____

2. _____

7. _____

6. _____

5. _____

4. _____

3. _____

The long tendons of the extensor muscles, associated with extension of the fingers and thumb, pass across the back of the wrist. One head of each of four dorsal interosseous muscles arises from each neighboring metacarpal, and they insert at the extensor tendons and base of the proximal phalanges of the 2nd–4th fingers; the extensor digiti minimi, extensor digitorum, and extensor carpi radialis brevis all arise at the lateral epicondyle of the humerus. The extensor digiti minimi inserts at the proximal phalanx of the 5th finger; the extensor digitorum inserts at the extensor expansions of the 2nd–5th fingers; and the extensor carpi radialis brevis at the base of the 2nd metacarpal bone. The abductor pollicis longus arises at the posterior of the proximal end of the radius and ulna and inserts at the base of the 1st metacarpal bone; the extensor pollicis brevis arises at the posterior of the radius and inserts at the base of the proximal phalanx of the thumb; the extensor pollicis longus arises at the posterior of the ulna and the interosseous membrane and inserts at the base of the distal phalanx of the thumb; and the extensor carpi radialis longus arises at the lateral supracondylar ridge of the humerus and inserts at the base of the 2nd metacarpal bone.

Answers

Muscles of the Lower Limb

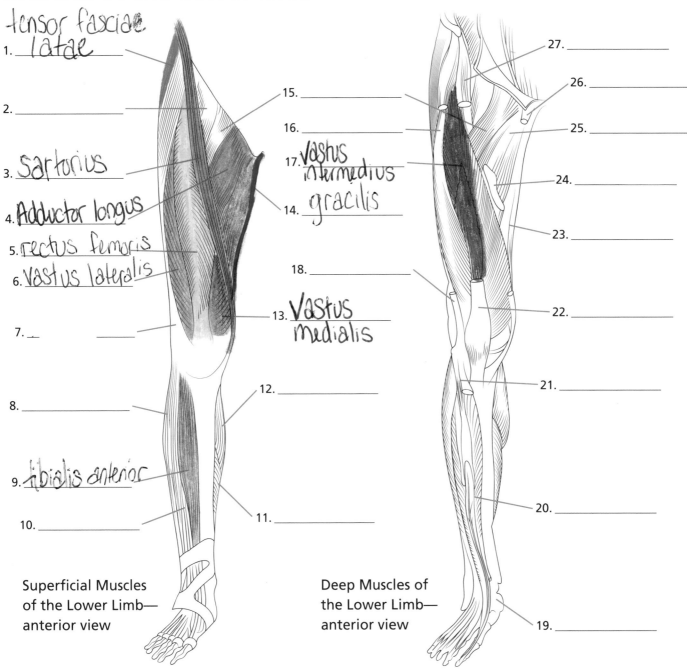

1. tensor fasciae latae
2. _____
3. sartorius
4. Adductor longus
5. rectus femoris
6. vastus lateralis
7. _____
8. _____
9. tibialis anterior
10. _____

15. _____
16. _____
17. vastus intermedius
14. gracilis
18. _____
13. vastus medialis
12. _____
11. _____

27. _____
26. _____
25. _____
24. _____
23. _____
22. _____
21. _____
20. _____
19. _____

**Superficial Muscles
of the Lower Limb—
anterior view**

**Deep Muscles of
the Lower Limb—
anterior view**

The superficial muscles of the lower limb are powerful and participate in locomotion. The quadriceps femoris muscle forms the major muscle mass of the front and outer side of the thigh. Its four parts are the rectus femoris, vastus lateralis, vastus medius, and vastus intermedius. The anterior compartment of the leg contains the muscles that move the foot upward (dorsiflex the ankle). The lateral compartment contains the fibularis longus and fibularis brevis muscles, which are responsible for turning the sole of the foot outward (eversion). The deep muscles of the thigh include the adductor muscles (adductor magnus, adductor longus, and adductor brevis). The muscles act to rotate, flex, and adduct the thigh. The extensor muscles (extensor hallucis longus and extensor digitorum longus) are the deep muscles of the leg, acting to extend the foot at the ankle (plantarflex the ankle) and the toes.

Answers

1. Tensor fascia lata, 2. Iliopsoas, 3. Sartorius, 4. Adductor longus, 5. Rectus femoris, 6. Vastus lateralis, 7. Iliotibial tract, 8. Fibularis longus, 9. Tibialis anterior, 10. Extensor digitorum longus, 11. Soleus, 12. Gastrocnemius, 13. Vastus medialis, 14. Gracilis, 15. Pectineus, 16. Vastus lateralis, 17. Vastus intermedius, 18. Iliotibial tract, 19. Tibialis anterior (cut), 20. Extensor hallucis longus, 21. Tibialis anterior (cut), 22. Rectus femoris (cut), 23. Adductor magnus, 24. Adductor longus (cut), 25. Adductor brevis, 26. Adductor longus (cut), 27. Sartorius (cut)

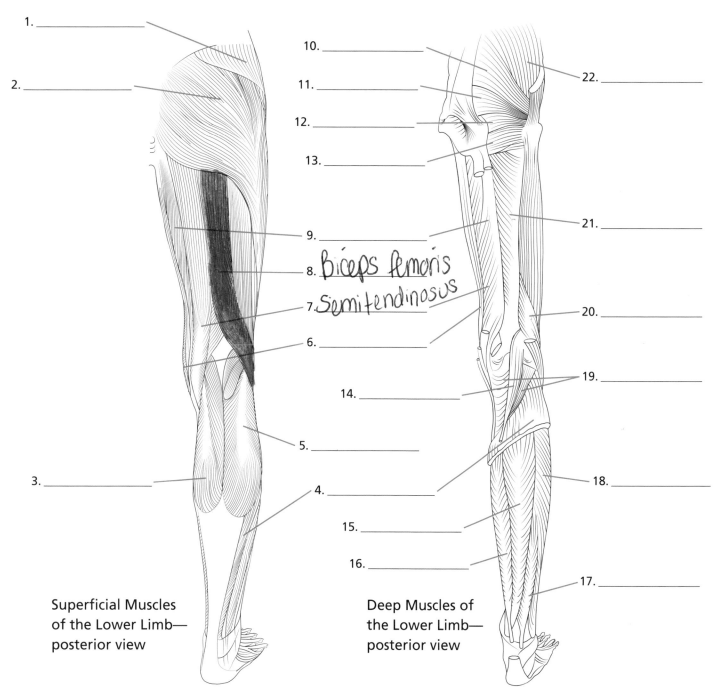

1. _____

2. _____

10. _____

11. _____

12. _____

13. _____

22. _____

21. _____

9. _____

8. *Biceps femoris*

7. *Semitendinosus*

6. _____

20. _____

14. _____

19. _____

5. _____

18. _____

3. _____

4. _____

15. _____

16. _____

17. _____

**Superficial Muscles
of the Lower Limb—
posterior view**

**Deep Muscles of
the Lower Limb—
posterior view**

Powerful superficial muscles surround and stabilize the hip region; the gluteal region is dominated by the gluteus maximus, the largest muscle in the body. The hamstring muscles in the posterior compartment of the thigh include the semimembranosus, semitendinosus, and the long head of the biceps femoris, which together extend the hip joint and flex the knee joint. The superficial group of the posterior compartment of the calf region contains the powerful gastrocnemius and soleus muscles, both critical for pushing off from the ground. The medial compartment of the thigh contains the adductor muscles, which pull the leg toward the midline. In the posterior compartment of the calf, the deep group of muscles pass behind the ankle joint and attach to the bones of the foot. The largest of these, the flexor hallucis longus, is critical to pushing off from the big toe during walking.

Answers

Muscles of the Lower Limb

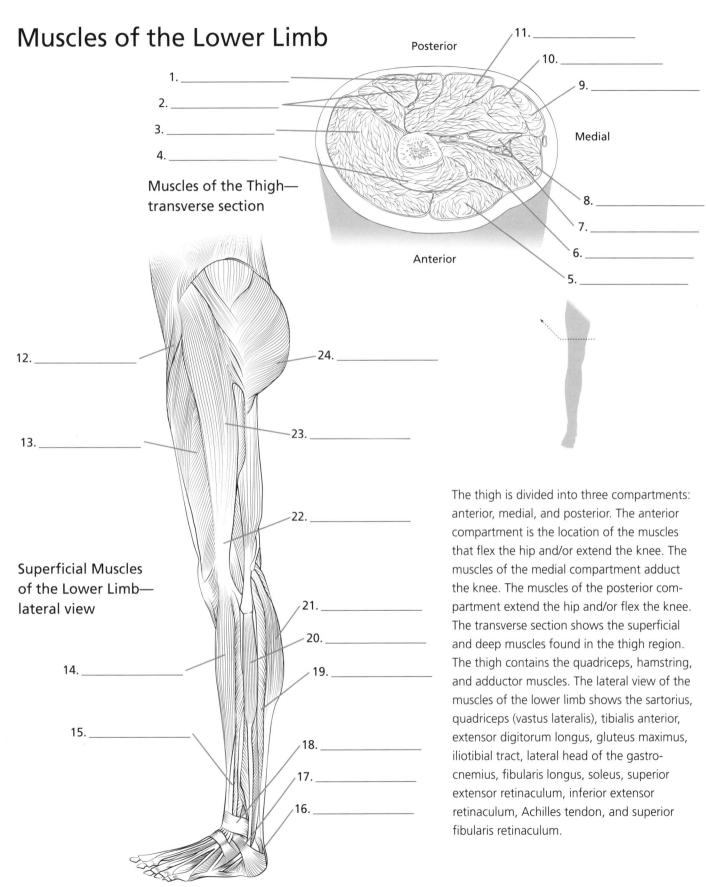

Posterior

11. _____

10. _____

9. _____

Medial

1. _____

2. _____

3. _____

4. _____

8. _____

7. _____

6. _____

5. _____

Anterior

Muscles of the Thigh—
transverse section

12. _____

13. _____

24. _____

23. _____

22. _____

Superficial Muscles
of the Lower Limb—
lateral view

21. _____

20. _____

19. _____

14. _____

15. _____

18. _____

17. _____

16. _____

The thigh is divided into three compartments: anterior, medial, and posterior. The anterior compartment is the location of the muscles that flex the hip and/or extend the knee. The muscles of the medial compartment adduct the knee. The muscles of the posterior compartment extend the hip and/or flex the knee. The transverse section shows the superficial and deep muscles found in the thigh region. The thigh contains the quadriceps, hamstring, and adductor muscles. The lateral view of the muscles of the lower limb shows the sartorius, quadriceps (vastus lateralis), tibialis anterior, extensor digitorum longus, gluteus maximus, iliotibial tract, lateral head of the gastrocnemius, fibularis longus, soleus, superior extensor retinaculum, inferior extensor retinaculum, Achilles tendon, and superior fibularis retinaculum.

Answers

1. Semitendinosus, 2. Biceps femoris, 3. Vastus lateralis, 4. Vastus intermedius, 5. Rectus femoris, 6. Vastus medialis, 7. Adductor longus, 8. Sartorius, 9. Gracilis, 10. Adductor magnus, 11. Semimembranosus, 12. Sartorius, 13. Quadriceps (vastus lateralis), 14. Tibialis anterior, 15. Extensor digitorum longus, 16. Achilles tendon, 17. Inferior extensor retinaculum, 18. Superior extensor retinaculum, 19. Soleus, 20. Fibularis longus, 21. Lateral head of gastrocnemius, 22. Iliotibial tract, 23. Tensor fascia lata, 24. Gluteus maximus

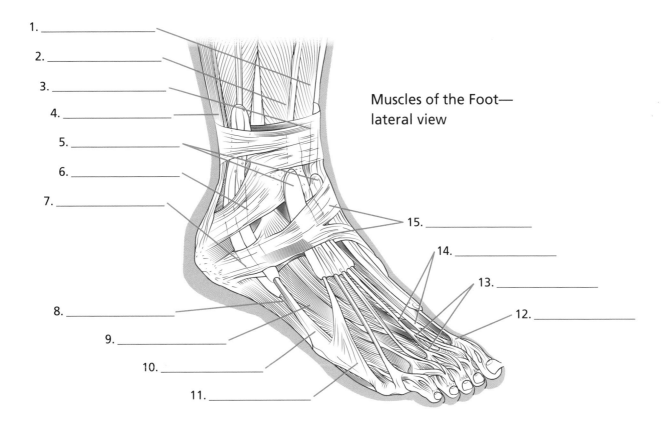

1. _____

2. _____

3. _____

4. _____

5. _____

6. _____

7. _____

Muscles of the Foot—
lateral view

15. _____

14. _____

13. _____

8. _____

12. _____

9. _____

10. _____

11. _____

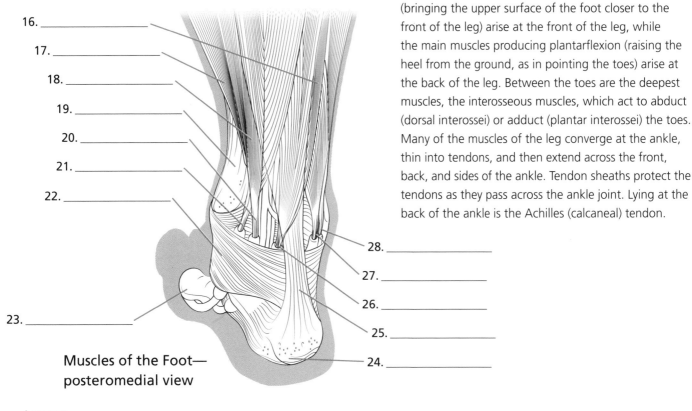

16. _____

17. _____

18. _____

19. _____

20. _____

21. _____

22. _____

23. _____

Muscles of the Foot—
posteromedial view

28. _____

27. _____

26. _____

25. _____

24. _____

The main muscles producing dorsiflexion of the foot (bringing the upper surface of the foot closer to the front of the leg) arise at the front of the leg, while the main muscles producing plantarflexion (raising the heel from the ground, as in pointing the toes) arise at the back of the leg. Between the toes are the deepest muscles, the interosseous muscles, which act to abduct (dorsal interossei) or adduct (plantar interossei) the toes. Many of the muscles of the leg converge at the ankle, thin into tendons, and then extend across the front, back, and sides of the ankle. Tendon sheaths protect the tendons as they pass across the ankle joint. Lying at the back of the ankle is the Achilles (calcaneal) tendon.

Answers

1. Extensor hallucis longus, 2. Extensor digitorum longus, 3. Superior extensor retinaculum, 4. Achilles (calcaneal) tendon, 5. Tendon sheaths, 6. Superior fibular retinaculum, 7. Inferior fibular retinaculum, 8. Fibularis longus tendon, 9. Extensor digitorum brevis, 10. Fibularis brevis tendon, 11. Fibularis tertius, 12. Extensor digitorum brevis tendons, 13. Extensor digitorum longus tendons, 14. Extensor digitorum longus tendons, 15. Inferior extensor retinaculum, 16. Flexor hallucis longus, 17. Tibialis posterior, 18. Flexor digitorum longus, 19. Tibia, 20. Flexor digitorum longus tendon, 21. Tibialis posterior tendon, 22. Flexor retinaculum, 23. First metatarsal, 24. Calcaneal tuberosity, 25. Achilles (calcaneal) tendon, 26. Flexor hallucis longus tendon, 27. Fibularis longus tendon, 28. Fibularis brevis tendon

Muscles of the Lower Limb

The muscles of the foot move the foot during walking, standing, running, and jumping. They also help to maintain the arches of the foot. The muscles are arranged in four layers. In the first (superficial) layer, the abductor hallucis flexes and abducts the big toe and supports the medial longitudinal arch. The flexor digitorum brevis attaches to and flexes the lateral four toes and supports the medial and lateral longitudinal arches. The abductor digiti minimi flexes and abducts the fifth toe and supports the lateral longitudinal arch. In the second, deeper layer are the flexor tendons that flex the toes: the flexor hallucis longus (the big toe) and the flexor digitorum longus (the other four lateral toes). The quadratus plantae (or flexor accessorius) also helps to flex the lateral four toes. The lumbricals help to extend the toes during walking and running.

6. _____

5. _____

4. _____

**First Layer
Muscles of the Foot—
inferior view**

1. _____

2. _____

13. _____

3. _____

7. _____

8. _____

12. _____

9. _____

10. _____

11. _____

**Second Layer
Muscles of the Foot—
inferior view**

Answers

1. Flexor digitorum brevis, 2. Abductor hallucis, 3. Fibrous flexor sheaths, 4. Abductor digiti minimi, 5. Plantar aponeurosis (cut), 6. Calcaneus, 7. Calcaneus, 8. Flexor digitorum longus tendon, 9. Flexor hallucis longus tendon, 10. Flexor hallucis brevis, 11. Lumbricals, 12. Flexor digiti minimi brevis, 13. Quadratus plantae

In the third layer is the flexor hallucis brevis, a short muscle that flexes the big toe and supports the medial longitudinal arch. There is adductor hallucis, the oblique head of which adducts the big toe and supports the transverse arch and the transverse head of which adducts the big toe. The flexor digiti minimi brevis flexes the little toe. In the fourth (deepest) layer of muscles, between the metatarsal bones of the foot, are the interossei. There are four dorsal interossei, which abduct the toes, and three plantar interossei, which adduct the toes.

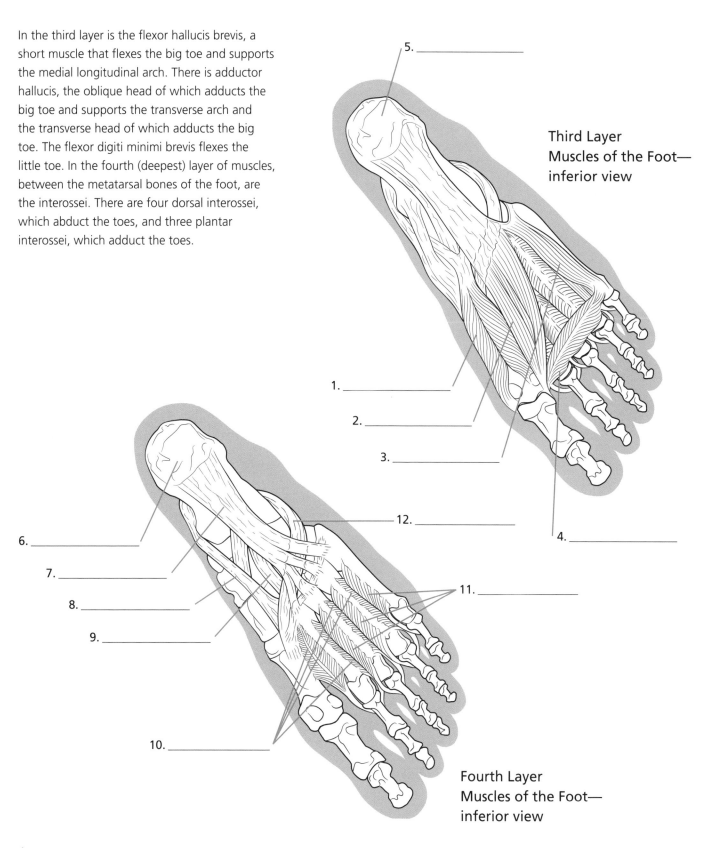

**Third Layer
Muscles of the Foot—
inferior view**

5. _____

1. _____

2. _____

3. _____

12. _____

4. _____

6. _____

7. _____

8. _____

9. _____

11. _____

10. _____

**Fourth Layer
Muscles of the Foot—
inferior view**

Answers

1. Flexor hallucis brevis, 2. Adductor hallucis (oblique head), 3. Flexor digiti minimi brevis, 4. Adductor hallucis (transverse head), 5. Calcaneus, 6. Calcaneus, 7. Long plantar ligament, 8. Calcaneonavicular ligament, 9. Calcaneocuboid ligament, 10. Dorsal interossei, 11. Plantar interossei, 12. Tendon of fibularis longus

Nervous System

Nervous System—
anterior view

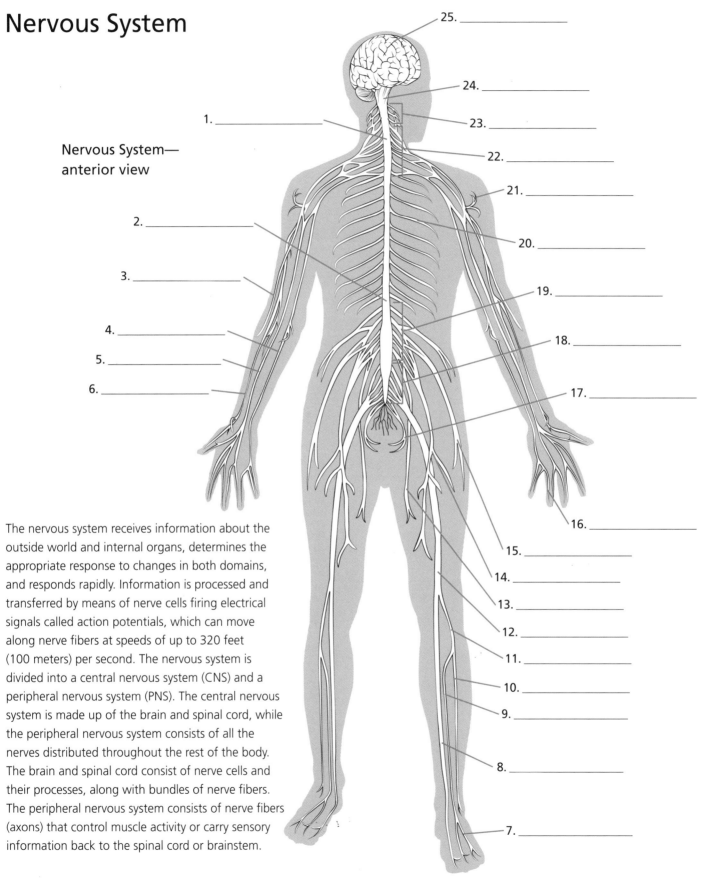

25. _____

24. _____

23. _____

22. _____

21. _____

20. _____

19. _____

18. _____

17. _____

16. _____

15. _____

14. _____

13. _____

12. _____

11. _____

10. _____

9. _____

8. _____

7. _____

1. _____

2. _____

3. _____

4. _____

5. _____

6. _____

The nervous system receives information about the outside world and internal organs, determines the appropriate response to changes in both domains, and responds rapidly. Information is processed and transferred by means of nerve cells firing electrical signals called action potentials, which can move along nerve fibers at speeds of up to 320 feet (100 meters) per second. The nervous system is divided into a central nervous system (CNS) and a peripheral nervous system (PNS). The central nervous system is made up of the brain and spinal cord, while the peripheral nervous system consists of all the nerves distributed throughout the rest of the body. The brain and spinal cord consist of nerve cells and their processes, along with bundles of nerve fibers. The peripheral nervous system consists of nerve fibers (axons) that control muscle activity or carry sensory information back to the spinal cord or brainstem.

Answers

1. Cervical enlargement of spinal cord, 2. Lumbosacral enlargement of spinal cord, 3. Musculocutaneous nerve, 4. Ulnar nerve, 5. Median nerve, 6. Radial nerve, 7. Plantar nerve, 8. Posterior tibial nerve, 9. Deep fibular nerve, 10. Superficial fibular nerve, 11. Common fibular nerve, 12. Sciatic nerve, 13. Obturator nerve, 14. Femoral nerve, 15. Lateral femoral cutaneous nerve, 16. Digital nerve, 17. Pudendal nerve, 18. Sacral plexus, 19. Lumbar plexus, 20. Intercostal nerve, 21. Axillary nerve, 22. Brachial plexus, 23. Cervical plexus, 24. Medulla oblongata, 25. Cerebral hemisphere

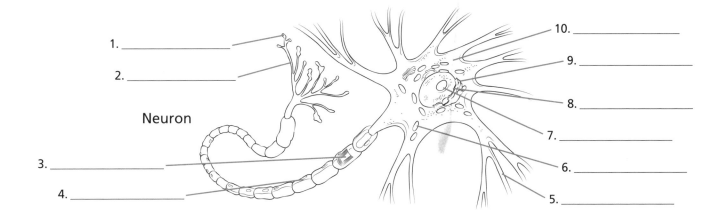

1. _____

2. _____

Neuron

3. _____

4. _____

10. _____

9. _____

8. _____

7. _____

6. _____

5. _____

Neurons are nerve cells that are specially designed to conduct nerve impulses to and from all parts of the body. A typical nerve cell has a cell body with a nucleus and a number of branching processes (dendrites), which receive incoming information, and an outgoing fiber (axon), which carries information away from the nerve cell body. Nerves connect the brain and the spinal cord (central nervous system) with peripheral regions such as the muscles, skin, and viscera (peripheral nervous system). The role of nerves is to carry signals that provide us with sensations or control our muscles and organs. The spinal cord is a cylindrical structure that occupies the vertebral canal. In the adult, the lower end of the spinal cord (conus medullaris) is usually located at the level of the L1 or L2 vertebra. Below this, the vertebral canal is occupied largely by the elongated rootlets of the lumbar and sacral spinal nerves (cauda equina).

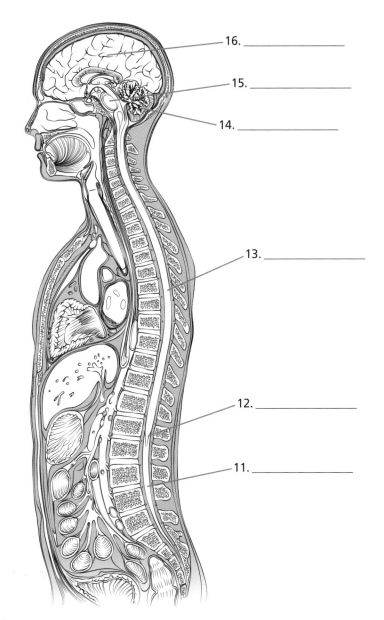

16. _____

15. _____

14. _____

13. _____

12. _____

11. _____

Central Nervous System—sagittal view

Answers

1. Synaptic knob, 2. Axon terminal, 3. Axon, 4. Myelin sheath, 5. Dendrite, 6. Mitochondria, 7. Nucleolus, 8. Nuclear membrane, 9. Golgi apparatus, 10. Cell body, 11. Region of cauda equina, 12. Conus medullaris, 13. Spinal cord, 14. Cerebellum, 15. Pons, 16. Cerebrum

Brain and Brain Functions

The brain lies within the cranial cavity of the skull and is made up of billions of nerve cells (neurons) and supporting cells (glia). The brain has gray matter (neurons and glia) and white matter (myelin). The human brain can be divided into four main parts: the cerebrum, diencephalon, brainstem, and cerebellum. The largest part is the cerebrum. It is formed by the cerebral hemispheres, which are joined together by a massive bundle of white matter (the corpus callosum), and it is covered by the gray matter of the cerebral cortex. The diencephalon lies beneath the cerebral hemispheres and has two main structures—the thalamus and the hypothalamus. The transverse section of the brain shows the gray matter of the cerebral cortex and the white matter that lies beneath the cortex. Forming the outer layer of the hemispheres of the brain, the cerebral cortex is a highly folded sheet of gray matter, with gyri folding outward and fissures and sulci folding inward. The meninges are three concentric protective membranes surrounding the brain and spinal cord. The outermost, the dura mater, is a tough fibrous layer. The middle membrane, the arachnoid mater, is a fragile network of collagen and elastin fibers. The innermost, the pia mater, is a delicate layer of collagen and elastic fibers containing many blood vessels.

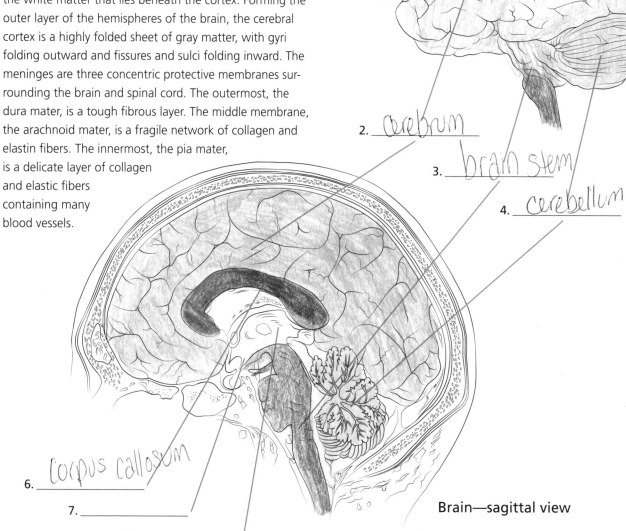

Brain—lateral view

1. _Gyri / gyrus_
5. _suci / sulcus_
2. _cerebrum_
3. _brain stem_
4. _cerebellum_
6. _corpus callosum_
7. _____
8. _____

Brain—sagittal view

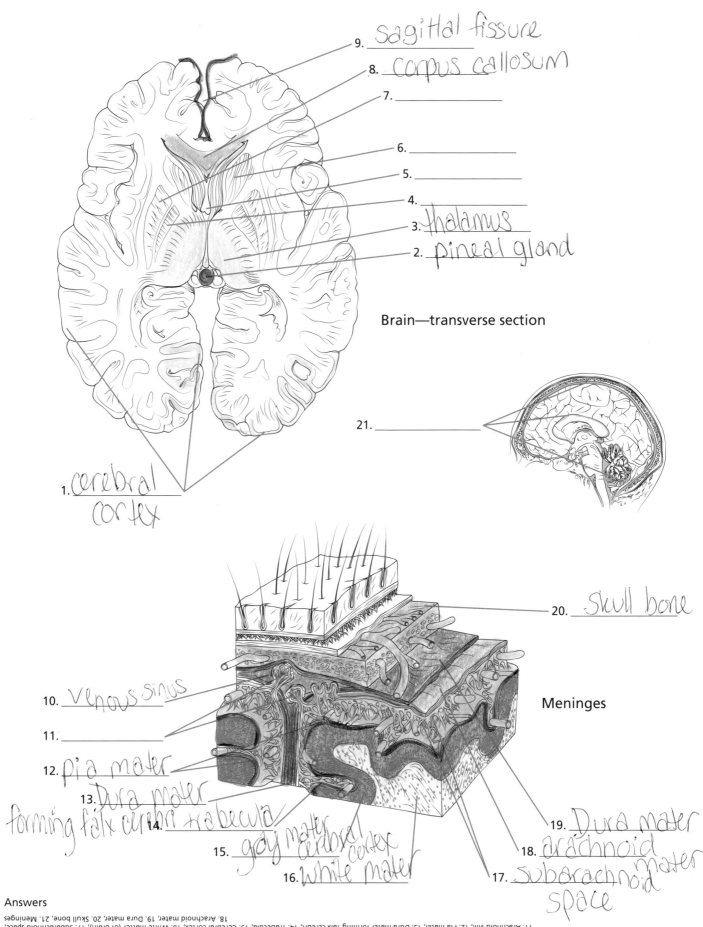

9. sagittal fissure
8. corpus callosum
7. _____
6. _____
5. _____
4. _____
3. thalamus
2. pineal gland

Brain—transverse section

21. _____

1. cerebral cortex

20. Skull bone

Meninges

10. venous sinus
11. _____
12. pia mater
13. Dura mater
14. forming falx cerebri trabecula
15. gray matter cerebral cortex
16. white mater

19. Dura mater
18. arachnoid mater
17. subarachnoid space

Answers

Brain and Brain Functions

The brain is divided longitudinally into two hemispheres by the sagittal fissure, with each hemisphere then divided into lobes marked by sulci and fissures in the folds of the cerebral cortex. The central sulcus separates the frontal lobe from the parietal lobe behind it. Located at the back of the brain is the occipital lobe. The temporal lobe lies beneath the frontal and parietal lobes. Each lobe is strongly linked, though not limited, to a particular function. The frontal lobe is involved in movement, thinking, behavior, and personality; the occipital lobe in the perception of vision; the parietal lobe in the perception of touch and comprehension of speech; and the temporal lobe in memory. The limbic system is a collective term for interconnected brain structures involved in behaviors associated with survival, including the expression of emotion, feeding, drinking, defense, and reproduction, as well as the formation of memory. The system's key components are the hippocampus, amygdala, septal area, and hypothalamus. The brain's four ventricles connect with each other, the central canal of the spinal cord, and the subarachnoid space surrounding the brain and spinal cord. The ventricles and subarachnoid space contain cerebrospinal fluid which acts as a shock absorber, cushioning the brain against mechanical forces.

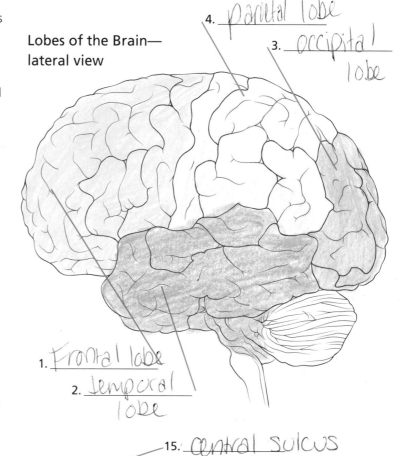

Lobes of the Brain—lateral view

4. parietal lobe
3. occipital lobe
1. Frontal lobe
2. Temporal lobe

Functional Areas of the Brain—lateral view

5. primary somatosensory cortex
6. primary motor cortex
7. Broca's speech
8. auditory association
9. primary auditory area

15. central sulcus
14. somatic sensory association area
13. visual association
12. primary visual area
11. reading comp. / visual association
10. wernicke's area

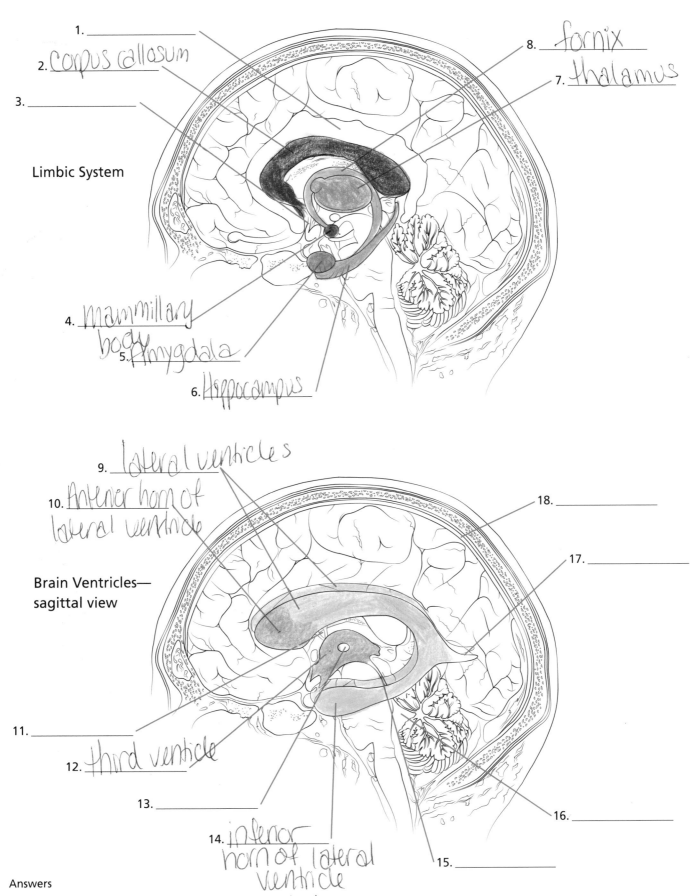

1. _____
2. corpus callosum
3. _____

Limbic System

8. fornix
7. thalamus

4. mammillary body
5. Amygdala
6. Hippocampus

9. lateral ventricles
10. Anterior horn of lateral ventricle

Brain Ventricles—
sagittal view

11. _____
12. third ventricle

13. _____

14. inferior horn of lateral ventricle

15. _____

16. _____

17. _____

18. _____

Answers

Cranial Nerves and Brainstem

There are twelve cranial nerves that lead directly from the brain and brainstem to muscles and sensory structures of the head and neck. Some of the cranial nerves also distribute nerves to the organs of the chest and the upper two-thirds of the gastrointestinal tract. Cranial nerve I is the olfactory nerve; cranial nerve II, the optic nerve; cranial nerve III, the oculomotor nerve; cranial nerve IV, the trochlear nerve; cranial nerve V, the trigeminal nerve, which has three sections (the ophthalmic, maxillary, and mandibular divisions); cranial nerve VI is the abducens nerve; cranial nerve VII, the facial nerve; cranial nerve VIII, the vestibulocochlear nerve; cranial nerve IX, the glossopharyngeal nerve; cranial nerve X, the vagus nerve; cranial nerve XI, the spinal accessory nerve; and cranial nerve XII, the hypoglossal nerve.

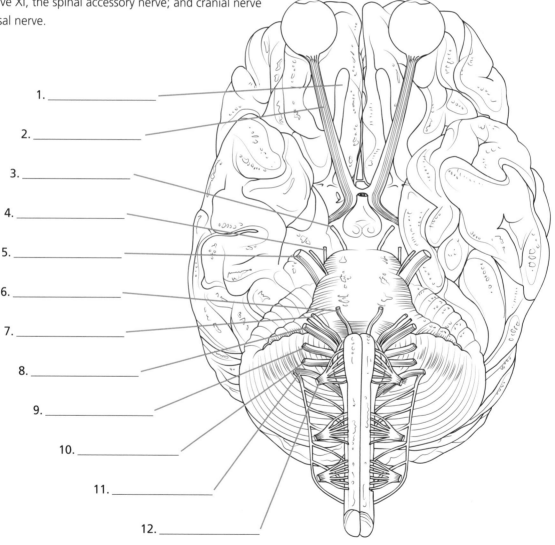

1. _____

2. _____

3. _____

4. _____

5. _____

6. _____

7. _____

8. _____

9. _____

10. _____

11. _____

12. _____

Cranial Nerves

Answers

The brainstem, which is continuous with the spinal cord below it, consists of the midbrain, pons, and medulla oblongata. Passing through the brainstem are ascending pathways, carrying sensory information from the spinal cord to the brain, and descending pathways, carrying motor signals down to the spinal cord. The oculomotor (cranial nerve III) and trochlear (IV) nerves arise in the midbrain. The pons lies below the midbrain. Lying below the pons is the medulla oblongata. The hypoglossal (XII), spinal accessory (XI), glossopharygeal (IX), and vagus (X) nerves arise in the medulla oblongata.

1. _____

2. _____

3. _____

4. _____

5. _____

6. _____

7. _____

19. _____

18. _____

17. _____

16. _____

15. _____

14. _____

13. _____

12. _____

11. _____

10. _____

9. _____

8. _____

Brainstem— posterior view

20. _____

21. _____

22. _____

23. _____

24. _____

25. _____

26. _____

27. _____

28. _____

29. _____

30. _____

31. _____

39. _____

38. _____

37. _____

36. _____

35. _____

34. _____

33. _____

32. _____

Brainstem— lateral view

Answers

1. Thalamus, 2. Choroid plexus of lateral ventricle, 3. Pineal body, 4. Cerebral peduncle, 5. Trochlear nerve (IV), 6. Cerebellar peduncles, 7. Atlas (C1), 8. Second cervical nerve, 9. Spinal accessory nerve (XI), 10. Sulcus limitans, 11. Facial colliculus, 12. Dorsal median sulcus, 13. Pons, 14. Inferior colliculus, 15. Superior colliculus, 16. Lateral geniculate body, 17. Medial geniculate body, 18. Pulvinar, 19. Habenula, 20. Choroid plexus, 21. Optic tract, 22. Optic nerve (II), 23. Cerebral peduncle, 24. Oculomotor nerve (III), 25. Abducens nerve (VI), 26. Facial nerve (VII), 27. Vestibulocochlear nerve (VIII), 28. Glossopharyngeal nerve (IX), 29. Hypoglossal nerve (XII), 30. Vagus nerve (X), 31. Spinal accessory nerve (XI), 32. Atlas (C1), 33. Medulla oblongata, 34. Middle cerebellar peduncle, 35. Trigeminal nerve (V), 36. Trochlear nerve (IV), 37. Inferior colliculus, 38. Lateral geniculate nucleus, 39. Superior colliculus

Spinal Cord

In an adult, the spinal cord extends from the base of the skull to a point about two-thirds of the way down the back, running through the vertebral canal. The cord itself has a central region of gray matter, which is divided into posterior (dorsal) and anterior (ventral) horns, and an intermediate zone. The gray matter is surrounded by white matter carrying ascending and descending fiber tracts. The dorsal horns are specialized to relay information up to the brain. The ventral horns contain motor neurons, which transmit messages out to the muscles via spinal nerves. The intermediate zone contains many interneurons involved in linking incoming sensory neurons with outgoing motor neurons to bring about automated (reflex) responses that do not involve the brain. The spinal cord has 31 pairs of nerves arising from it.

16. dorsal funiculus

15. spinal gray matter

1. spinothalamic tract

2. posterior spinal artery

3. anterior/ventral horns

4. dorsal rootlets

14. dorsal horns

13. spinal ganglion

5. posterior ramus of a spinal nerves

6. anterior ramus of a spinal nerve

7. pia mater

8. arachnoid mater

9. dura mater

10. ventral rootlets

11. axon

12. myelin sheath of schwann cell

Spinal Cord—cross-sectional view

Answers

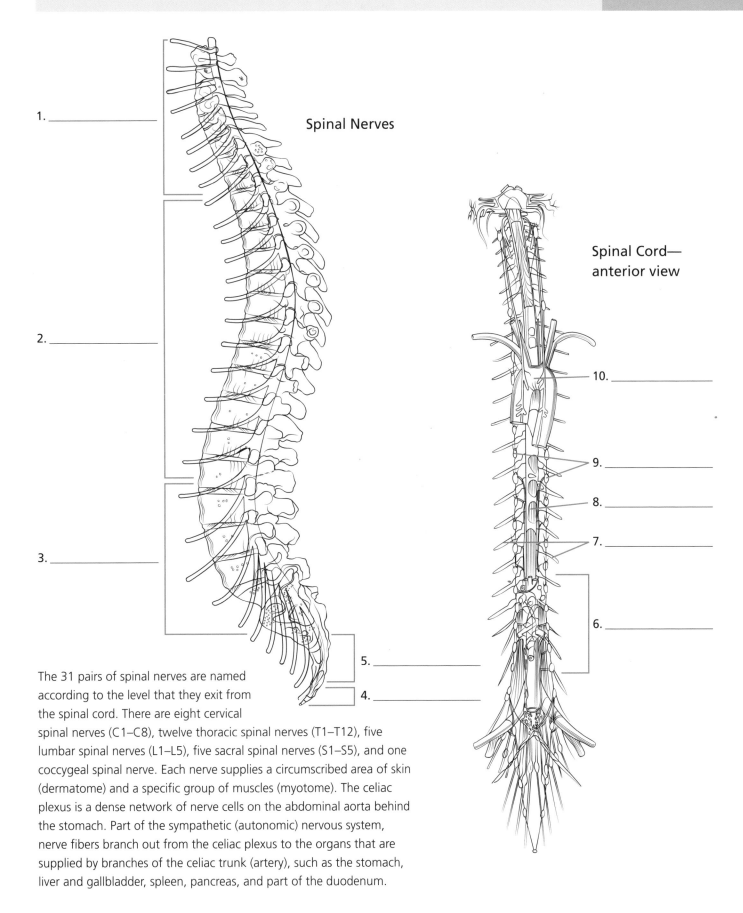

Spinal Nerves

Spinal Cord—
anterior view

1. _____

2. _____

3. _____

4. _____

5. _____

6. _____

7. _____

8. _____

9. _____

10. _____

The 31 pairs of spinal nerves are named
according to the level that they exit from
the spinal cord. There are eight cervical
spinal nerves (C1–C8), twelve thoracic spinal nerves (T1–T12), five
lumbar spinal nerves (L1–L5), five sacral spinal nerves (S1–S5), and one
coccygeal spinal nerve. Each nerve supplies a circumscribed area of skin
(dermatome) and a specific group of muscles (myotome). The celiac
plexus is a dense network of nerve cells on the abdominal aorta behind
the stomach. Part of the sympathetic (autonomic) nervous system,
nerve fibers branch out from the celiac plexus to the organs that are
supplied by branches of the celiac trunk (artery), such as the stomach,
liver and gallbladder, spleen, pancreas, and part of the duodenum.

Answers

1. Spinal nerves C1–C8, 2. Spinal nerves T1–T12, 3. Spinal nerves L1–L5, 4. Coccygeal spinal nerve, 5. Spinal nerves S1–S5,
6. Celiac (solar) plexus, 7. Peripheral nerves, 8. Spinal cord, 9. Sympathetic ganglia, 10. Aortic arch

Dermatomes

Dermatomes are the areas of skin supplied by a single spinal nerve; however, there is usually some overlap between adjacent dermatomes. Each of the 31 segments of the spinal cord gives rise to a pair of spinal nerves, which carry sensory messages into and motor messages out of the central nervous system. These spinal nerves branch into and service particular areas of the body. Spinal nerve C1 does not play a role in dermatomes. The skin of the face is supplied by branches of the trigeminal nerve (cranial nerves V1, V2, and V3).

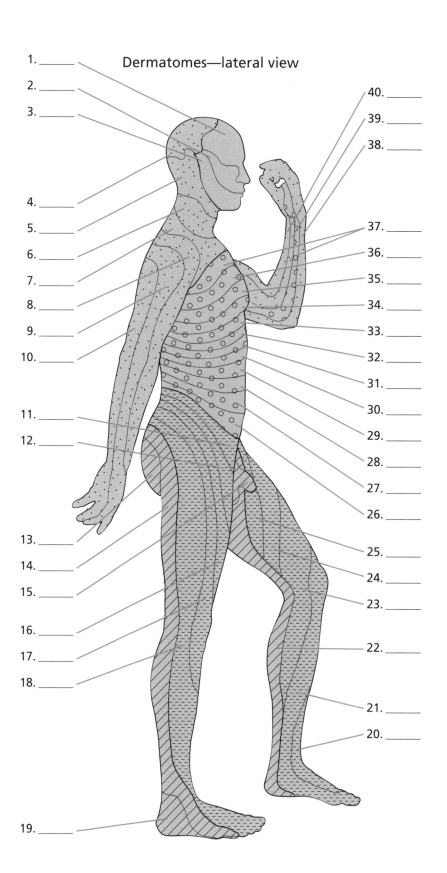

Dermatomes—lateral view

1. _____
2. _____
3. _____
4. _____
5. _____
6. _____
7. _____
8. _____
9. _____
10. _____
11. _____
12. _____
13. _____
14. _____
15. _____
16. _____
17. _____
18. _____
19. _____
20. _____
21. _____
22. _____
23. _____
24. _____
25. _____
26. _____
27. _____
28. _____
29. _____
30. _____
31. _____
32. _____
33. _____
34. _____
35. _____
36. _____
37. _____
38. _____
39. _____
40. _____

Answers

1. V1, 2. V2, 3. V3, 4. C2, 5. C3, 6. C4, 7. C5, 8. C6, 9. C7, 10. C8, 11. L1, 12. L2, 13. L2, 14. S1, 15. S3, 16. L1, 17. L4, 18. L5, 19. S2, 20. L5, 21. L4, 22. S2, 23. L3, 24. L2, 25. L1, 26. T12, 27. T11, 28. T10, 29. T9, 30. T8, 31. T7, 32. T6, 33. T5, 34. T4, 35. T3, 36. T2, 37. T1, 38. C8, 39. C5, 40. C6

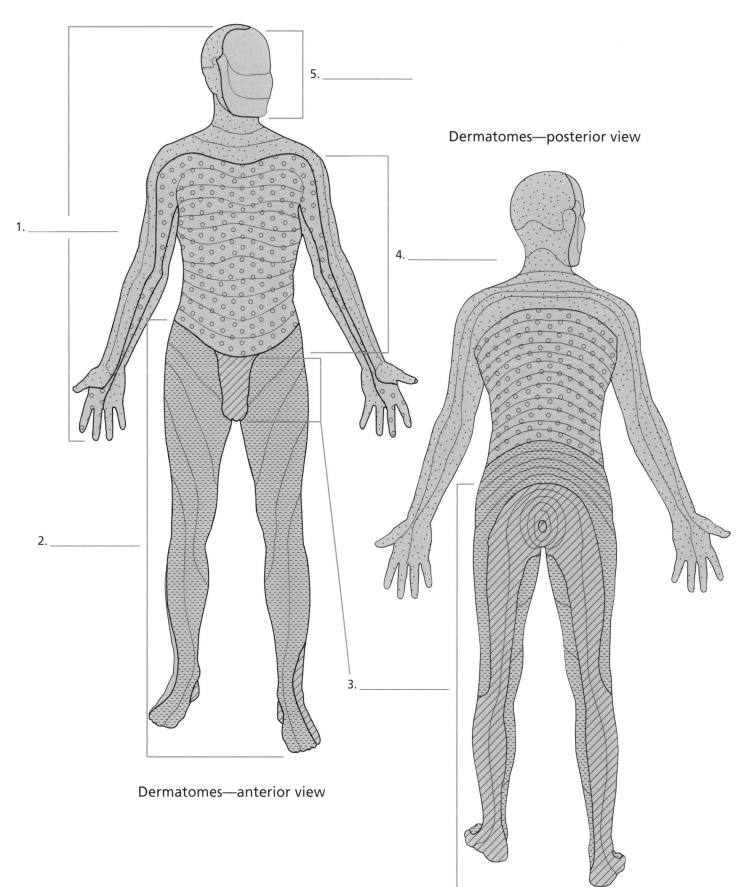

5. _____

Dermatomes—posterior view

4. _____

1. _____

2. _____

3. _____

Dermatomes—anterior view

Answers

1. Dermatomes C2–C8, 2. Dermatomes L1–L5, 3. Dermatomes S1–S5, 4. Dermatomes T1–T12, 5. Dermatomes V1–V3

Nerves of the Head and Neck

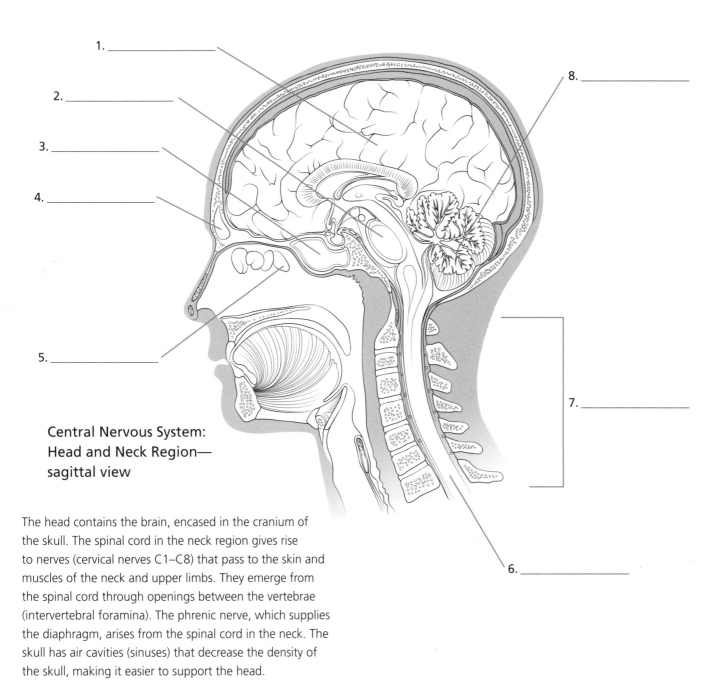

1. _____

2. _____

3. _____

4. _____

5. _____

6. _____

7. _____

8. _____

**Central Nervous System:
Head and Neck Region—
sagittal view**

The head contains the brain, encased in the cranium of the skull. The spinal cord in the neck region gives rise to nerves (cervical nerves C1–C8) that pass to the skin and muscles of the neck and upper limbs. They emerge from the spinal cord through openings between the vertebrae (intervertebral foramina). The phrenic nerve, which supplies the diaphragm, arises from the spinal cord in the neck. The skull has air cavities (sinuses) that decrease the density of the skull, making it easier to support the head.

Answers

1. Cerebrum, 2. Pons , 3. Sphenoid sinus, 4. Frontal sinus, 5. Ethmoidal sinuses, 6. Spinal cord, 7. Cervical vertebrae (C1–C7), 8. Cerebellum

Nerves of the Head and Neck— lateral view

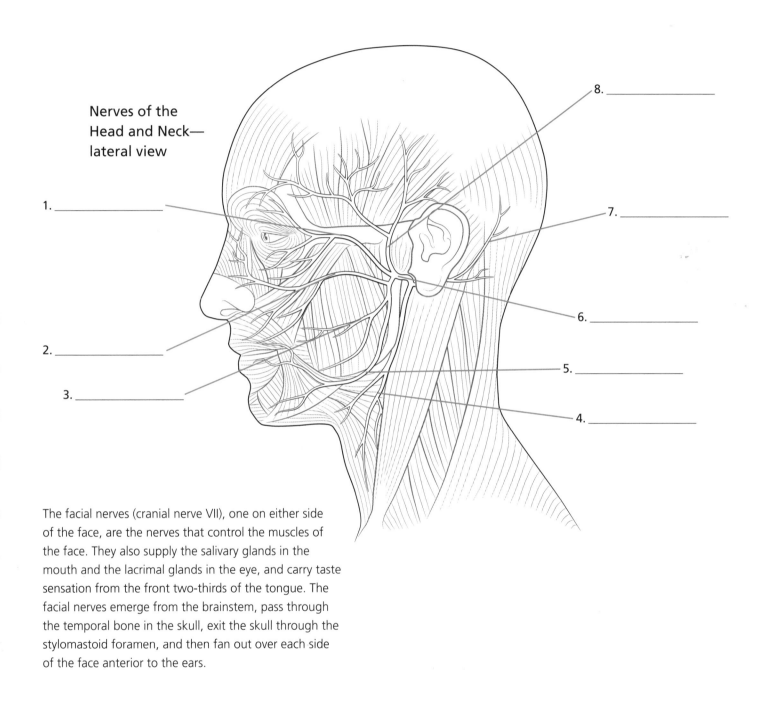

1. _____

2. _____

3. _____

8. _____

7. _____

6. _____

5. _____

4. _____

The facial nerves (cranial nerve VII), one on either side of the face, are the nerves that control the muscles of the face. They also supply the salivary glands in the mouth and the lacrimal glands in the eye, and carry taste sensation from the front two-thirds of the tongue. The facial nerves emerge from the brainstem, pass through the temporal bone in the skull, exit the skull through the stylomastoid foramen, and then fan out over each side of the face anterior to the ears.

Major Nerves of the Upper and Lower Limbs

The major nerves serving the upper limb are the ulnar nerve, the radial nerve, and the median musculocutaneous nerve. These nerves arise at the brachial plexus in the shoulder region. The ulnar nerve innervates the flexor carpi ulnaris muscle, half of the flexor digitorum profundus muscle, and small muscles of the hand. It provides sensation to the skin on the palm side of the little finger side of the hand. The radial nerve innervates the extensor muscles in the back of the arm and supplies the skin over the back of the arm and hand on the thumb side. The median nerve innervates some of the flexor muscles of the forearm, some of the muscles of the thumb, and some of the skin on the front of the hand. The musculocutaneous nerve innervates the flexor muscles of the arm.

Major Nerves of
the Upper Limb—
posterior view

9. _____

1. _____

2. _____

3. _____

4. _____

8. _____

7. _____

5. _____

6. _____

Major Nerves
of the Wrist—
palmar view

10. _____

11. _____

12. _____

13. _____

14. _____

15. _____

Major Nerves of
the Upper Limb—
anterior view

Answers

1. Radial nerve, 2. Median nerve, 3. Ulnar nerve, 4. Musculocutaneous nerve, 5. Anterior interosseus nerve, 6. Digital nerves of radial nerve, 7. Superficial branch of radial nerve,
8. Deep branch of radial nerve, 9. Axillary nerve, 10. Common palmar digital branches of median nerve, 11. Superficial branch of ulnar nerve, 12. Flexor retinaculum, 13. Median nerve,
14. Superficial branch of radial nerve, 15. Ulnar nerve

Major Nerves of the Lower Limb— anterior view

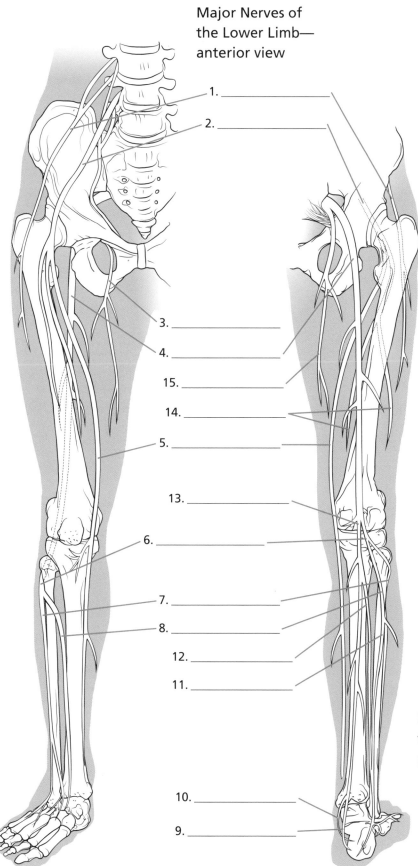

1. _____

2. _____

3. _____

4. _____

15. _____

14. _____

5. _____

13. _____

6. _____

7. _____

8. _____

12. _____

11. _____

10. _____

9. _____

The femoral nerve, the obturator nerve, and the sciatic nerve are the principal nerves of the lower limb, supplying the muscles and much of the skin. The femoral and obturator nerves arise at the lumbar plexus, and the sciatic nerve arises at the sacral plexus. The sciatic nerve and its branches (the tibial and common fibular nerves) supply the hamstring muscles of the thigh and all the muscles of the leg and foot. The obturator nerve supplies the gracilis muscle and the adductor muscles of the thigh and the skin on the medial part of the thigh. The femoral nerve and its major branch, the saphenous nerve, supply the muscles at the front of the thigh, the joints of the hip and knee, and the skin of the thigh, leg, and foot.

Major Nerves of the Lower Limb— posterior view

Answers

1. Lateral femoral cutaneous nerve, 2. Femoral nerve, 3. Obturator nerve, 4. Sciatic nerve, 5. Saphenous nerve, 6. Common fibular nerve, 7. Superficial fibular nerve, 8. Deep fibular nerve, 9. Lateral plantar nerve, 10. Medial plantar nerve, 11. Lateral sural cutaneous nerve, 12. Medial sural cutaneous nerve, 13. Tibial nerve, 14. Branches from femoral nerve, 15. Posterior femoral cutaneous nerve

Autonomic Nervous System

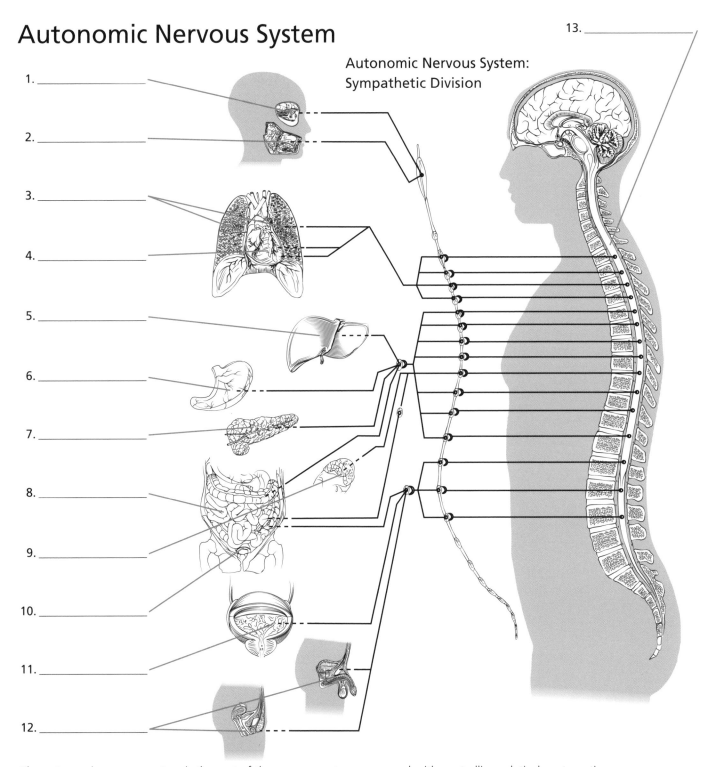

Autonomic Nervous System:
Sympathetic Division

1. _____

2. _____

3. _____

4. _____

5. _____

6. _____

7. _____

8. _____

9. _____

10. _____

11. _____

12. _____

13. _____

The autonomic nervous system is the part of the nervous system concerned with controlling relatively automatic body functions. It is divided into sympathetic and parasympathetic divisions. Often referred to as the "fight-or-flight" system, the sympathetic division comes into play during emergency situations. In such situations, the sympathetic division reacts by increasing heart rate, blood pressure, and blood sugar levels; the pupils dilate; the breathing becomes more rapid; and blood flow to the muscles is increased. The sympathetic division also has an important effect on the control of body temperature and on the urinary and reproductive organs.

Answers

1. Eye and lacrimal glands, 2. Salivary glands, 3. Lungs, 4. Heart, 5. Liver, 6. Stomach, 7. Pancreas, 8. Small intestine, 9. Adrenal medulla, 10. Large intestine and rectum, 11. Bladder, 12. Reproductive organs, 13. Spinal cord

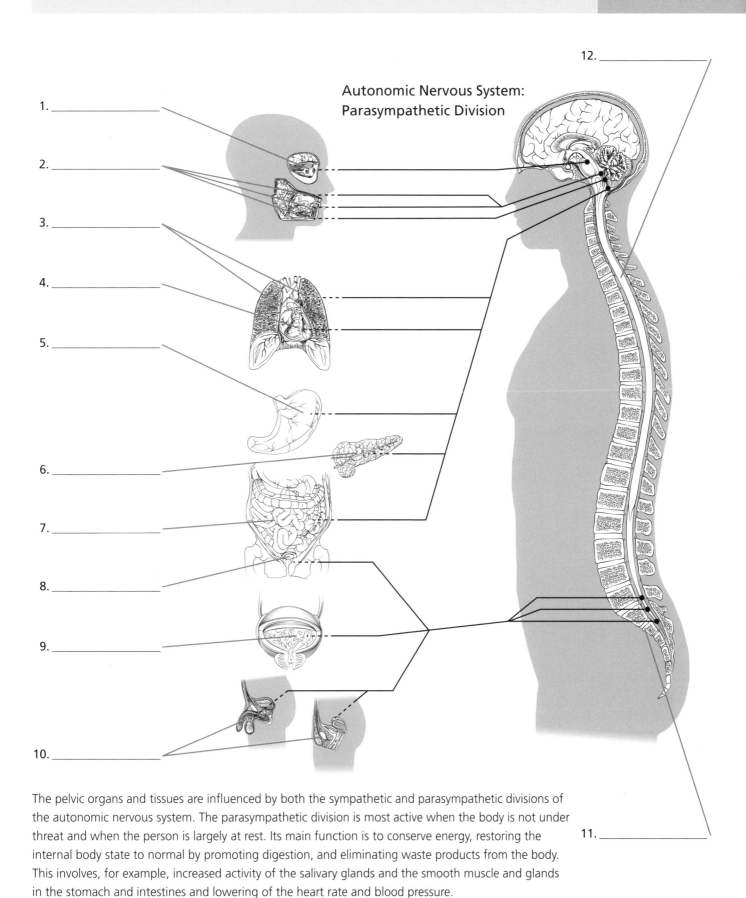

Autonomic Nervous System: Parasympathetic Division

1. _____

2. _____

3. _____

4. _____

5. _____

6. _____

7. _____

8. _____

9. _____

10. _____

11. _____

12. _____

The pelvic organs and tissues are influenced by both the sympathetic and parasympathetic divisions of the autonomic nervous system. The parasympathetic division is most active when the body is not under threat and when the person is largely at rest. Its main function is to conserve energy, restoring the internal body state to normal by promoting digestion, and eliminating waste products from the body. This involves, for example, increased activity of the salivary glands and the smooth muscle and glands in the stomach and intestines and lowering of the heart rate and blood pressure.

Answers

1. Eye, 2. Salivary glands, 3. Lungs, 4. Heart, 5. Stomach, 6. Pancreas, 7. Small intestine, 8. Rectum, 9. Bladder, 10. Reproductive organs, 11. Sacrum, 12. Spinal cord

Sight

Eye—anterior view

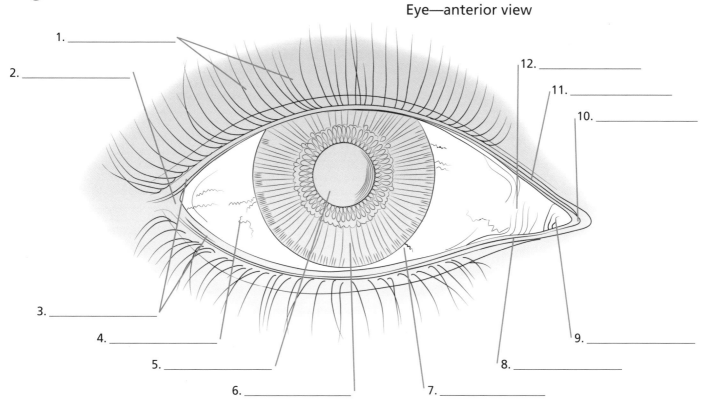

1. _____

2. _____

12. _____

11. _____

10. _____

3. _____

4. _____

5. _____

6. _____

7. _____

8. _____

9. _____

The eye is the organ of sight. Light rays entering the eye strike the cornea, which refracts the rays, bringing them closer together. The rays then pass through the lens, which focuses them on the back of the retina. The retina consists of a layer of light-sensitive cells—rods and cones. When stimulated by light, the rods and cones send electrical signals along the cells of the optic nerve. The optic nerve connects the eye to the brain. The impulses travel via the optic nerves to the occipital cortex and other parts of the brain, where they are interpreted as vision. The eyelids (palpebrae) protect the eyes. Each eyelid is covered on the outside by the skin and on the inside by the conjunctiva. The eyelid margin is lined with a few rows of eyelashes (cilia) and has openings for numerous glands.

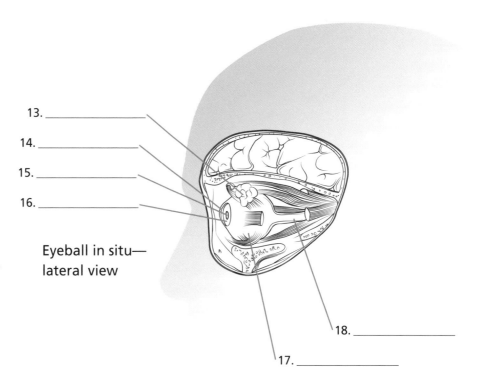

13. _____

14. _____

15. _____

16. _____

Eyeball in situ— lateral view

18. _____

17. _____

Answers

Eye—lateral view

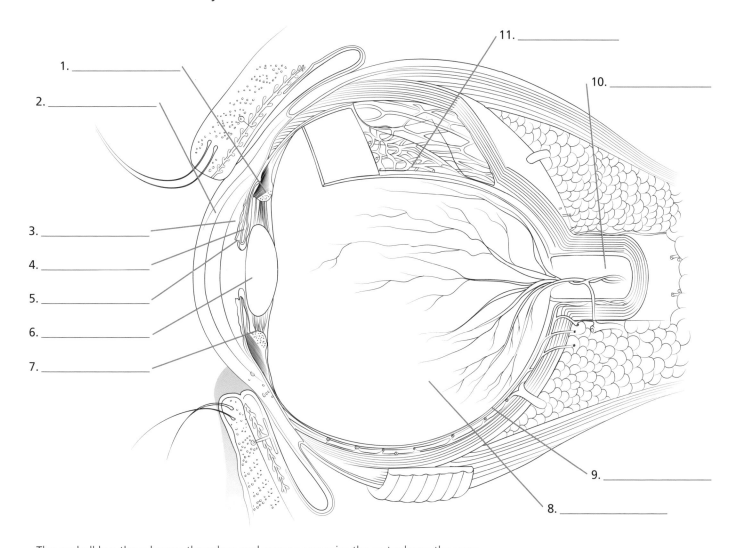

1. _____
2. _____
3. _____
4. _____
5. _____
6. _____
7. _____
8. _____
9. _____
10. _____
11. _____

The eyeball has three layers: the sclera and cornea comprise the outer layer, the uvea and lens comprise the middle layer, and the retina is the inner layer. The sclera gives the eyeball its spherical shape, with the front of the sphere covered by the curved layer of the transparent cornea. The front of the sclera is covered by a thin membrane called the conjunctiva. The uvea has three parts: the choroid, ciliary body, and iris. The choroid is the posterior part of the uvea, the ciliary body contains muscles, and the iris is the anterior part, lying behind the cornea. The retina lines the inside of the back part of the eyeball. The nerve fibers from the retina converge at the optic nerve. The eyeball is divided into two fluid-filled cavities. The anterior cavity is in front of the lens and is made up of the anterior and posterior chambers on the front and sides of the iris, respectively. The anterior cavity contains aqueous humor, and the posterior cavity contains the vitreous body (a gelatinlike mass).

Answers

1. Ciliary body, 2. Cornea, 3. Anterior chamber (of anterior cavity), 4. Posterior chamber (of anterior cavity), 5. Iris, 6. Lens, 7. Ciliary fibers, 8. Vitreous body (of posterior cavity), 9. Retina, 10. Optic nerve, 11. Choroid

Hearing and Balance

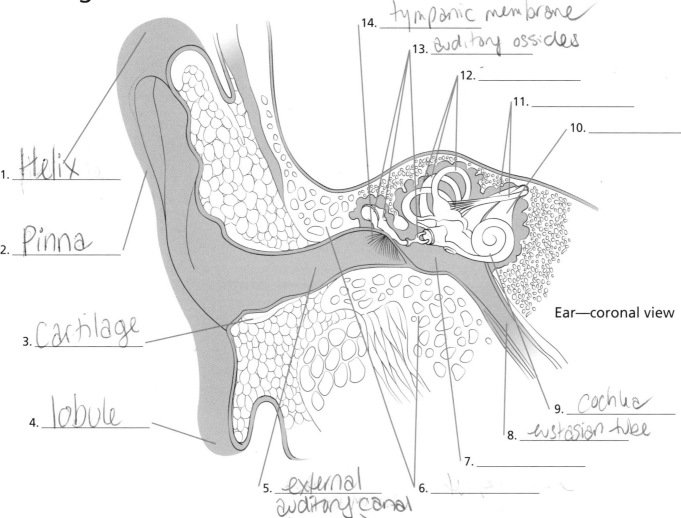

1. Helix

2. Pinna

3. Cartilage

4. lobule

5. external auditory canal (meatus)

6. _____

7. _____

8. eustasian tube

9. cochlea

10. _____

11. _____

12. _____

13. auditory ossicles

14. tympanic membrane

Ear—coronal view

Tympanic Membrane (eardrum)—internal view

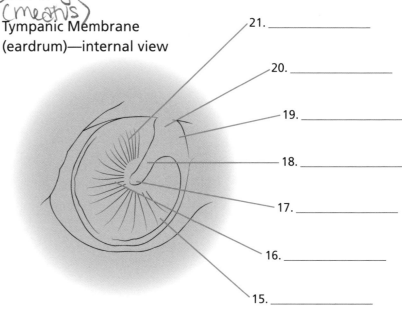

15. _____

16. _____

17. _____

18. _____

19. _____

20. _____

21. _____

Sound is conducted to the brain through the ear, the organ of hearing. The outer ear acts as a funnel to collect sound waves and guide them to the tympanic membrane. When sound waves strike the tympanic membrane, they cause it to vibrate and pass vibrations to the ossicles in the middle ear. The ossicles transmit vibrations into the cochlea. There they disturb the cochlear fluid, exciting tiny hair cells in the organ of Corti that send nerve impulses along the cochlear branch of the vestibulocochlear nerve to the hearing center in the temporal lobe of the brain, where sounds are interpreted. The ear also is the organ of balance; head movements are detected deep inside the inner ear by the semicircular ducts.

Answers

1. Helix, 2. Pinna, 3. Cartilage, 4. Lobule, 5. External auditory canal (meatus), 6. Temporal bone, 7. Middle ear (tympanic cavity), 8. Eustachian (auditory) tube, 9. Cochlea, 10. Cochlear nerve branch, 11. Vestibular nerve branches, 12. Semicircular canals, 13. Ossicles (malleus, incus, and stapes), 14. Tympanic membrane (eardrum), 15. Cone of light, 16. Pars tensa, 17. Umbo, 18. Handle of malleus, 19. Anterior malleolar fold, 20. Lateral process of malleus, 21. Posterior malleolar fold

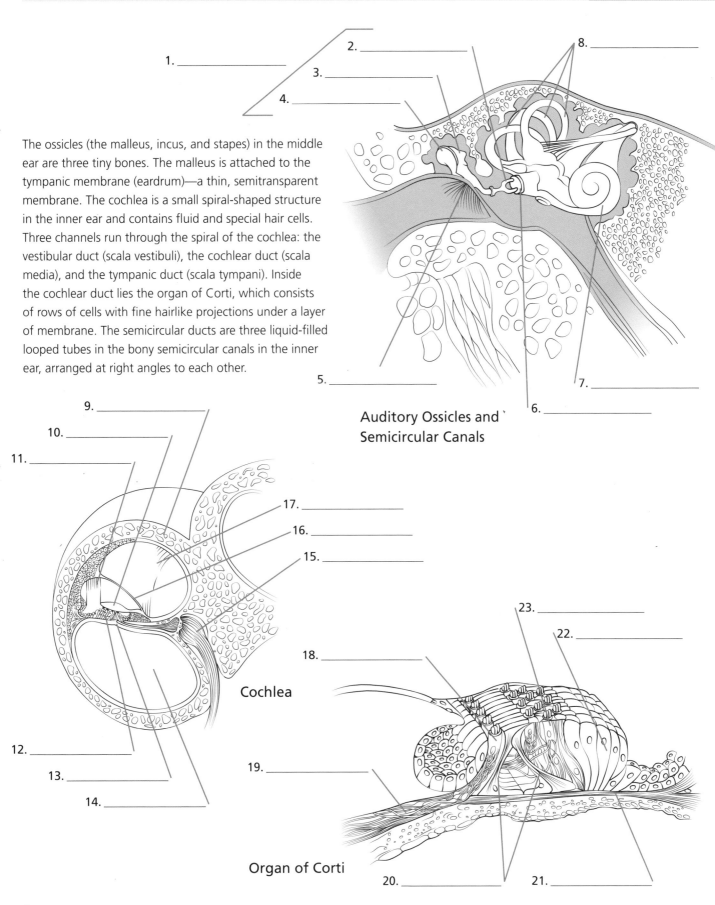

1. _____

2. _____

3. _____

4. _____

8. _____

The ossicles (the malleus, incus, and stapes) in the middle ear are three tiny bones. The malleus is attached to the tympanic membrane (eardrum)—a thin, semitransparent membrane. The cochlea is a small spiral-shaped structure in the inner ear and contains fluid and special hair cells. Three channels run through the spiral of the cochlea: the vestibular duct (scala vestibuli), the cochlear duct (scala media), and the tympanic duct (scala tympani). Inside the cochlear duct lies the organ of Corti, which consists of rows of cells with fine hairlike projections under a layer of membrane. The semicircular ducts are three liquid-filled looped tubes in the bony semicircular canals in the inner ear, arranged at right angles to each other.

5. _____

7. _____

6. _____

Auditory Ossicles and Semicircular Canals

9. _____

10. _____

11. _____

17. _____

16. _____

15. _____

12. _____

13. _____

14. _____

Cochlea

23. _____

22. _____

18. _____

19. _____

20. _____

21. _____

Organ of Corti

Answers

1. Ossicles, 2. Stapes, 3. Incus, 4. Malleus, 5. Eardrum, 6. Stapes footplate covering vestibular vestibule (oval) window, 7. Cochlea, 8. Semicircular canals, 9. Bony cochlear wall, 10. Tectorial membrane, 11. Cochlear duct (scala media), 12. Organ of Corti, 13. Tympanic duct (scala tympani), 14. Basilar membrane, 15. Cochlear nerve, 16. Vestibular membrane, 17. Vestibular duct (scala vestibuli), 18. Inner hair cell, 19. Nerve fibers, 20. Pillar cells, 21. Basilar membrane, 22. Phalangeal cell, 23. Outer hair cell

Taste and Smell

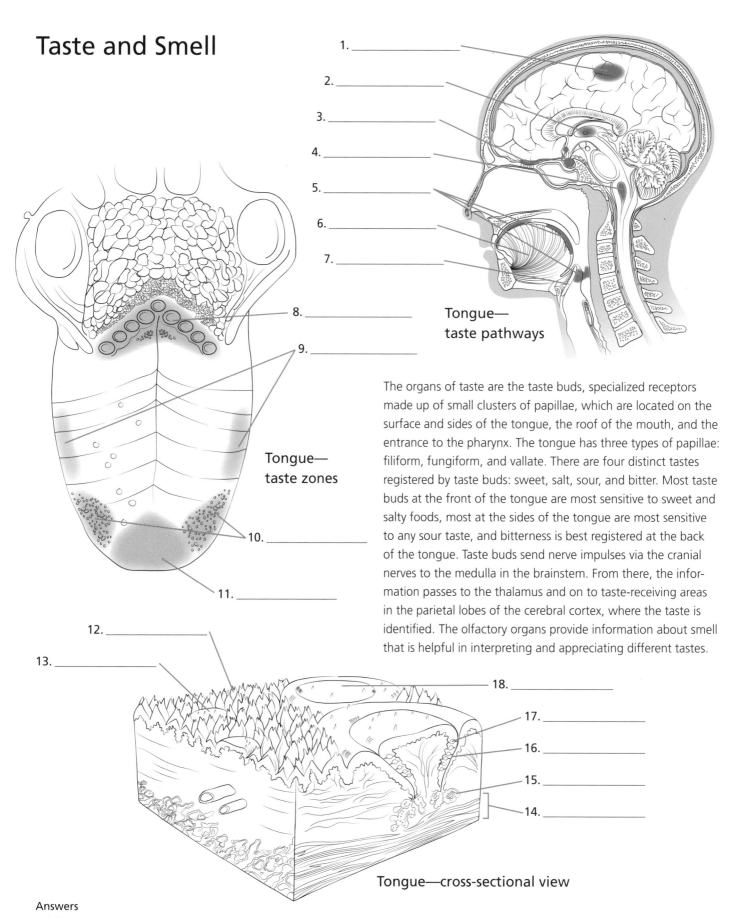

1. _____

2. _____

3. _____

4. _____

5. _____

6. _____

7. _____

**Tongue—
taste pathways**

8. _____

9. _____

**Tongue—
taste zones**

10. _____

11. _____

12. _____

13. _____

18. _____

17. _____

16. _____

15. _____

14. _____

Tongue—cross-sectional view

The organs of taste are the taste buds, specialized receptors made up of small clusters of papillae, which are located on the surface and sides of the tongue, the roof of the mouth, and the entrance to the pharynx. The tongue has three types of papillae: filiform, fungiform, and vallate. There are four distinct tastes registered by taste buds: sweet, salt, sour, and bitter. Most taste buds at the front of the tongue are most sensitive to sweet and salty foods, most at the sides of the tongue are most sensitive to any sour taste, and bitterness is best registered at the back of the tongue. Taste buds send nerve impulses via the cranial nerves to the medulla in the brainstem. From there, the information passes to the thalamus and on to taste-receiving areas in the parietal lobes of the cerebral cortex, where the taste is identified. The olfactory organs provide information about smell that is helpful in interpreting and appreciating different tastes.

Answers

1. Parietal lobe, 2. Thalamus, 3. Olfactory bulbs, 4. Medulla oblongata, 5. Regions of taste buds, 6. Tongue, 7. Epiglottis, 8. Bitter taste zone, 9. Sour taste zone, 10. Salt taste zone, 11. Sweet taste zone, 12. Filiform papilla, 13. Fungiform papilla, 14. Muscular layer, 15. Serous gland (Ebner's gland), 16. Trench, 17. Taste bud, 18. Vallate papilla

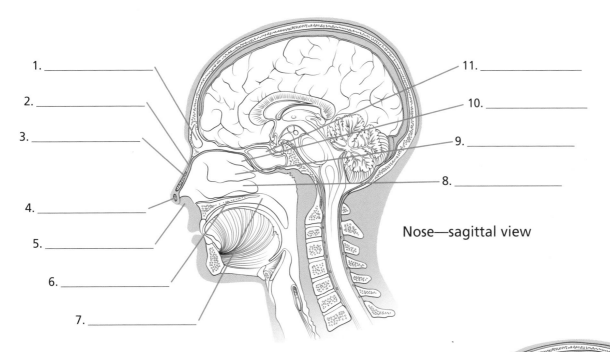

1. _____

2. _____

3. _____

4. _____

5. _____

6. _____

7. _____

11. _____

10. _____

9. _____

8. _____

Nose—sagittal view

The olfactory system is concerned with the sense of smell. The receptors for the sense of smell are situated in the olfactory mucosa that line the upper part of the cavities of the nose. A small area (about 1 square inch [5 square centimeters]) of mucous membrane (the olfactory epithelium) contains about twenty nerves served by about 100 million olfactory receptor cells. The olfactory nerves pass through foramina in the cribriform plate of the ethmoid bone, terminating in the olfactory bulbs in the brain. The olfactory bulbs are ovoid structures forming forward extensions of the olfactory area of the brain. Sensations received by the olfactory receptor cells are trans-mitted to the olfactory bulb and then sent along the olfactory tract to the medial olfactory area or lateral olfactory area in the brain.

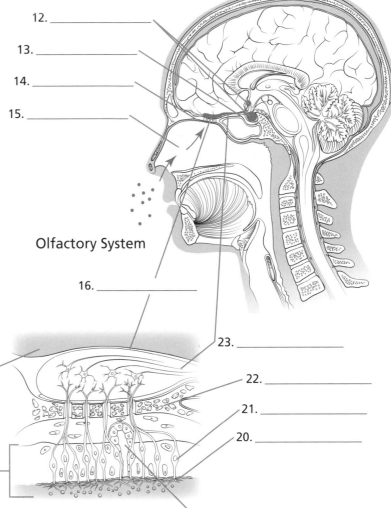

12. _____

13. _____

14. _____

15. _____

Olfactory System

16. _____

17. _____

18. _____

23. _____

22. _____

21. _____

20. _____

19. _____

Answers

Lymphatic System

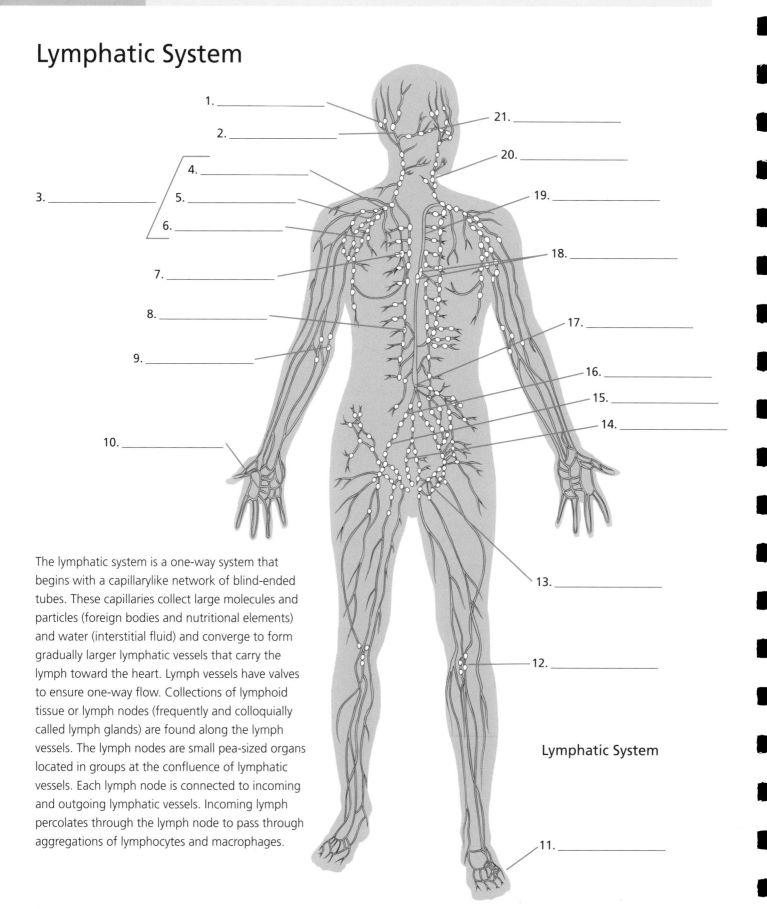

1. _____

2. _____

3. _____

4. _____

5. _____

6. _____

7. _____

8. _____

9. _____

10. _____

11. _____

12. _____

13. _____

14. _____

15. _____

16. _____

17. _____

18. _____

19. _____

20. _____

21. _____

The lymphatic system is a one-way system that begins with a capillarylike network of blind-ended tubes. These capillaries collect large molecules and particles (foreign bodies and nutritional elements) and water (interstitial fluid) and converge to form gradually larger lymphatic vessels that carry the lymph toward the heart. Lymph vessels have valves to ensure one-way flow. Collections of lymphoid tissue or lymph nodes (frequently and colloquially called lymph glands) are found along the lymph vessels. The lymph nodes are small pea-sized organs located in groups at the confluence of lymphatic vessels. Each lymph node is connected to incoming and outgoing lymphatic vessels. Incoming lymph percolates through the lymph node to pass through aggregations of lymphocytes and macrophages.

Lymphatic System

Answers

1. Retroauricular nodes, 2. Parotid nodes, 3. Axillary nodes, 4. Apical axillary nodes, 5. Lateral group, 6. Anterior group, 7. Parasternal nodes, 8. Posterior intercostal nodes, 9. Cubital nodes, 10. Palmar and dorsal plexus, 11. Plantar and dorsal plexus, 12. Popliteal nodes (posterior side of the knee), 13. Inguinal and femoral nodes, 14. Internal iliac nodes, 15. External iliac nodes, 16. Common iliac nodes, 17. Cisterna chyli, 18. Posterior mediastinal nodes, 19. Thoracic duct, 20. Cervical nodes, 21. Buccal nodes

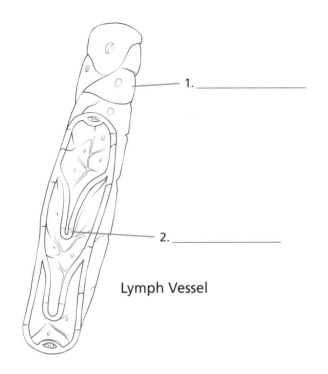

1. _____

2. _____

Lymph Vessel

Lymph is a fluid that is filtered and cleaned by white blood cells in the lymph nodes (glands). A lymph node consists of a mass of lymphatic tissue that is surrounded by a fibrous capsule. The nodes cluster in groups along the lymphatic vessels. Each node is enclosed in a fibrous capsule and has compartments divided by partitions of collagen fibers (trabeculae). Each lymph node is connected to incoming and outgoing lymph vessels. Afferent lymphatic vessels bring lymph fluid to the node from surrounding tissues; efferent vessels transport lymph to the veins. Numerous valves prevent the backflow of lymph as it passes through the lymphatic vessels.

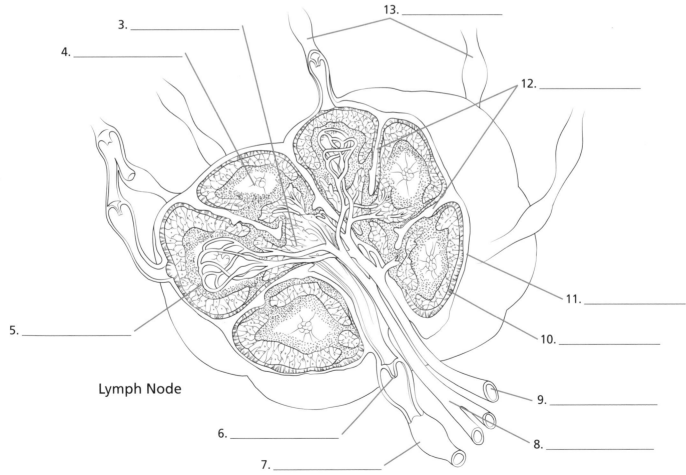

3. _____

4. _____

13. _____

12. _____

11. _____

10. _____

5. _____

9. _____

8. _____

6. _____

7. _____

Lymph Node

Answers

Lymphoid Organs

Lymphoid organs include mucosa-associated lymphoid tissue, the thymus, and the spleen. Concentrations of lymphoid tissue are found in the tonsils and in the gut. The tonsils are arranged around the entrance to the respiratory and digestive tracts to protect the body from bacteria and viruses that may enter from the mouth and nose. The tonsils lie under the surface lining of the mouth and throat and produce lymphocytes that cross into the mouth and throat. There are three sets of tonsils, named according to their positions. The lingual tonsil lies on the back third of the tongue; the palatine tonsils lie on either side of the back of the tongue, between pillars of tissue that join the soft palate to the tongue; and the pharyngeal or nasopharyngeal tonsils (adenoids) lie in the space behind the nose.

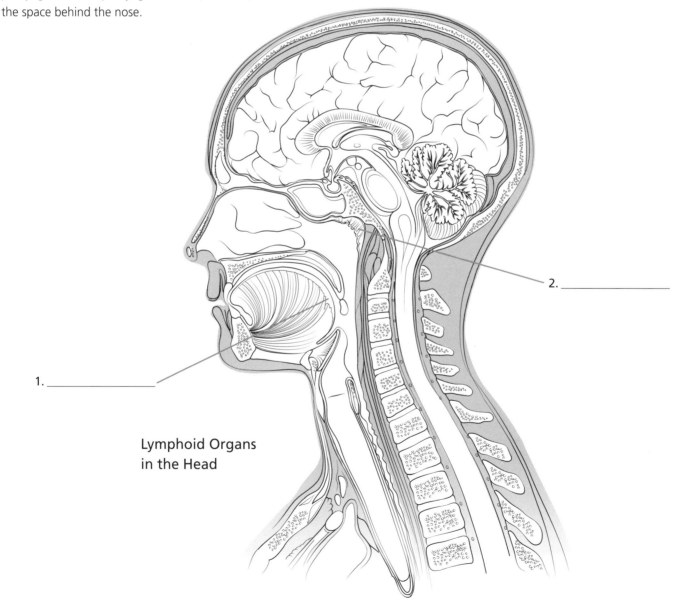

1. _____

2. _____

**Lymphoid Organs
in the Head**

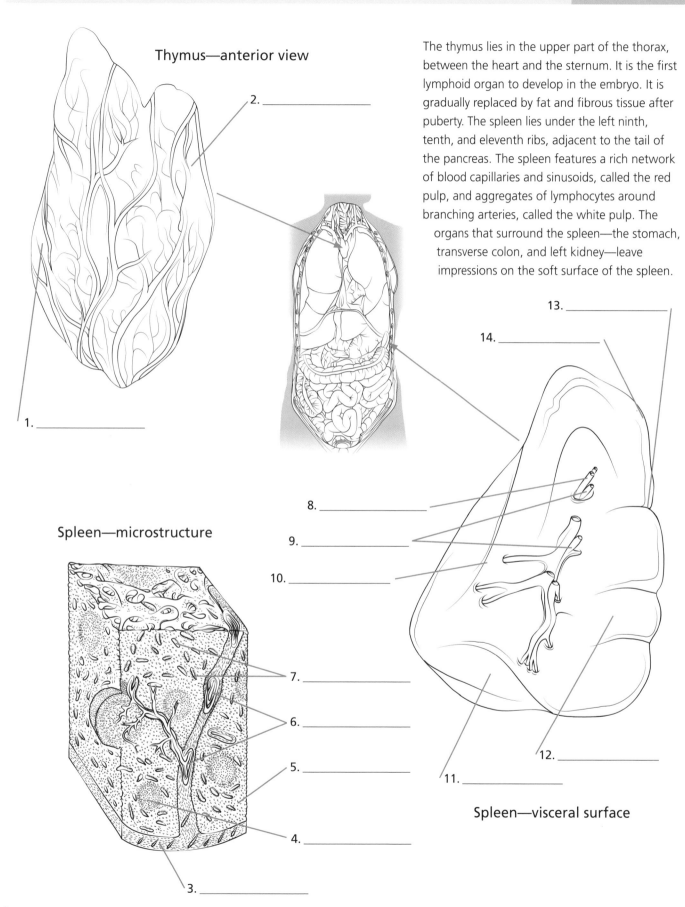

Thymus—anterior view

2. _____

1. _____

The thymus lies in the upper part of the thorax, between the heart and the sternum. It is the first lymphoid organ to develop in the embryo. It is gradually replaced by fat and fibrous tissue after puberty. The spleen lies under the left ninth, tenth, and eleventh ribs, adjacent to the tail of the pancreas. The spleen features a rich network of blood capillaries and sinusoids, called the red pulp, and aggregates of lymphocytes around branching arteries, called the white pulp. The organs that surround the spleen—the stomach, transverse colon, and left kidney—leave impressions on the soft surface of the spleen.

13. _____

14. _____

8. _____

9. _____

10. _____

12. _____

11. _____

Spleen—visceral surface

Spleen—microstructure

7. _____

6. _____

5. _____

4. _____

3. _____

Answers

Circulatory System

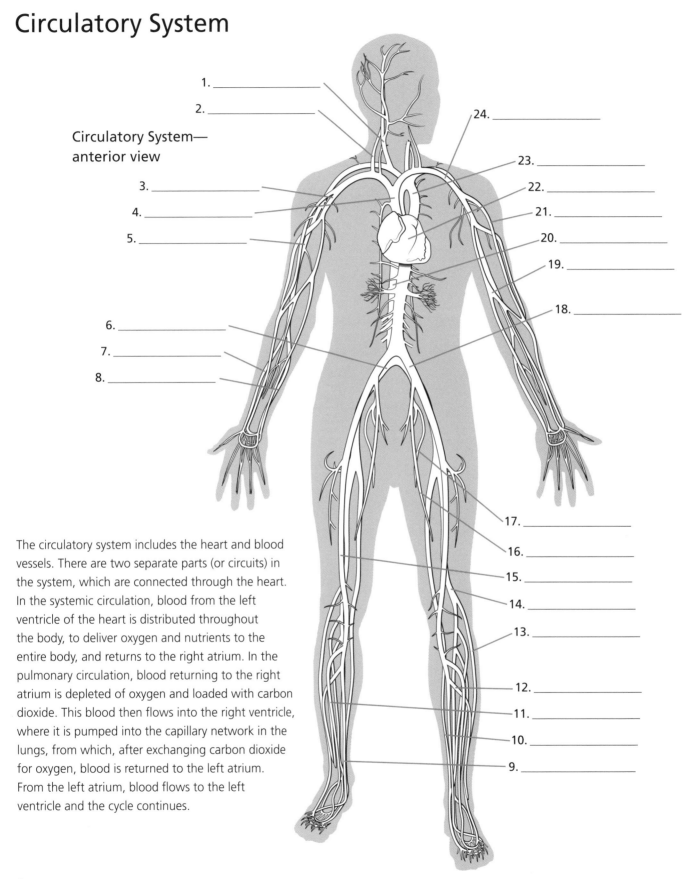

Circulatory System—
anterior view

1. _____

2. _____

3. _____

4. _____

5. _____

6. _____

7. _____

8. _____

24. _____

23. _____

22. _____

21. _____

20. _____

19. _____

18. _____

17. _____

16. _____

15. _____

14. _____

13. _____

12. _____

11. _____

10. _____

9. _____

The circulatory system includes the heart and blood vessels. There are two separate parts (or circuits) in the system, which are connected through the heart. In the systemic circulation, blood from the left ventricle of the heart is distributed throughout the body, to deliver oxygen and nutrients to the entire body, and returns to the right atrium. In the pulmonary circulation, blood returning to the right atrium is depleted of oxygen and loaded with carbon dioxide. This blood then flows into the right ventricle, where it is pumped into the capillary network in the lungs, from which, after exchanging carbon dioxide for oxygen, blood is returned to the left atrium. From the left atrium, blood flows to the left ventricle and the cycle continues.

Answers

1. Common carotid artery, 2. External jugular vein, 3. Axillary vein, 4. Superior vena cava, 5. Brachial artery, 6. Common iliac artery, 7. Radial artery, 8. Ulnar artery, 9. Posterior tibial artery, 10. Great saphenous vein, 11. Anterior tibial artery, 12. Fibular artery, 13. Small saphenous vein, 14. Popliteal vein, 15. Femoral artery, 16. Obturator artery, 17. Obturator vein, 18. Common iliac artery, 19. Basilic vein, 20. Inferior vena cava, 21. Cephalic vein, 22. Heart, 23. Arch of aorta, 24. Subclavian vein

Blood—cellular level

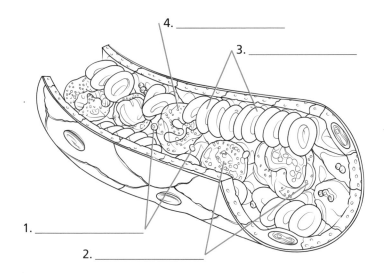

4. _____

3. _____

1. _____

2. _____

Blood is composed of red blood cells (erythrocytes), various types of white blood cells (leukocytes), and platelets in a solution of water, electrolytes, and proteins called plasma. Red blood cells carry oxygen to the tissues, while plasma carries essential nutrients to the tissues. White blood cells are the agents of the body's immune system. Blood vessels form an intricate system through which the blood circulates. As they conduct oxygenated blood away from the heart to the tissues, arteries branch into arterioles, which become capillaries, allowing oxygen and other nutrients to be exchanged in the surrounding tissues. Then, as the capillaries conduct deoxygenated blood back to the heart, they increase in diameter; capillaries become venules, which join to become veins. Many veins contain one-way valves. Fenestrated capillaries have openings, or windows, in their walls; continuous capillaries do not.

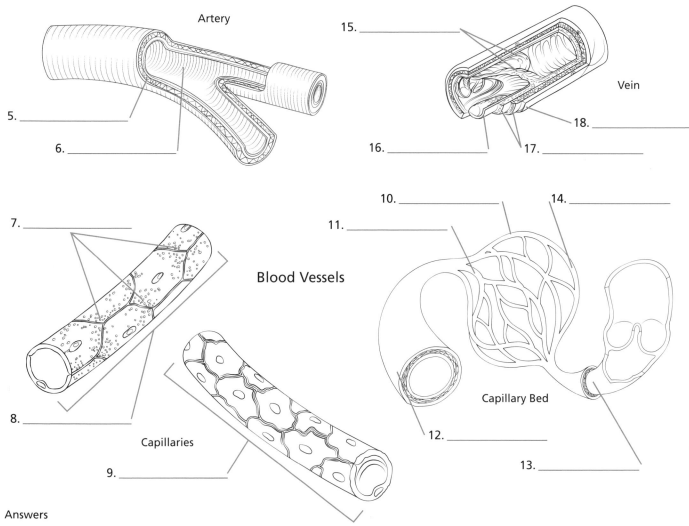

Artery

5. _____

6. _____

15. _____

Vein

18. _____

16. _____

17. _____

7. _____

Blood Vessels

8. _____

Capillaries

9. _____

10. _____

11. _____

14. _____

Capillary Bed

12. _____

13. _____

Answers

Circulatory System

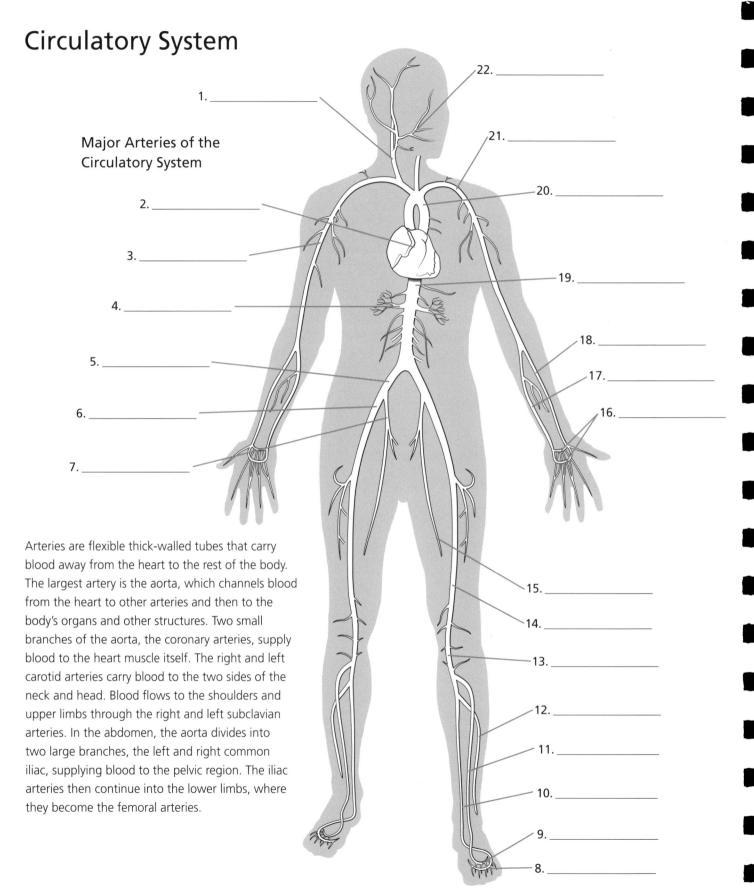

Major Arteries of the Circulatory System

1. _____
2. _____
3. _____
4. _____
5. _____
6. _____
7. _____
8. _____
9. _____
10. _____
11. _____
12. _____
13. _____
14. _____
15. _____
16. _____
17. _____
18. _____
19. _____
20. _____
21. _____
22. _____

Arteries are flexible thick-walled tubes that carry blood away from the heart to the rest of the body. The largest artery is the aorta, which channels blood from the heart to other arteries and then to the body's organs and other structures. Two small branches of the aorta, the coronary arteries, supply blood to the heart muscle itself. The right and left carotid arteries carry blood to the two sides of the neck and head. Blood flows to the shoulders and upper limbs through the right and left subclavian arteries. In the abdomen, the aorta divides into two large branches, the left and right common iliac, supplying blood to the pelvic region. The iliac arteries then continue into the lower limbs, where they become the femoral arteries.

Answers

1. Common carotid artery, 2. Heart, 3. Brachial artery, 4. Renal artery, 5. Common iliac artery, 6. External iliac artery, 7. Internal iliac artery, 8. Plantar arch, 9. Arcuate artery, 10. Posterior tibial artery, 11. Anterior tibial artery, 12. Fibular artery, 13. Popliteal artery, 14. Obturator artery, 15. Femoral artery, 16. Palmar arches, 17. Ulnar artery, 18. Radial artery, 19. Abdominal aorta, 20. Arch of aorta, 21. Axillary artery, 22. Facial artery

Major Veins of the Circulatory System

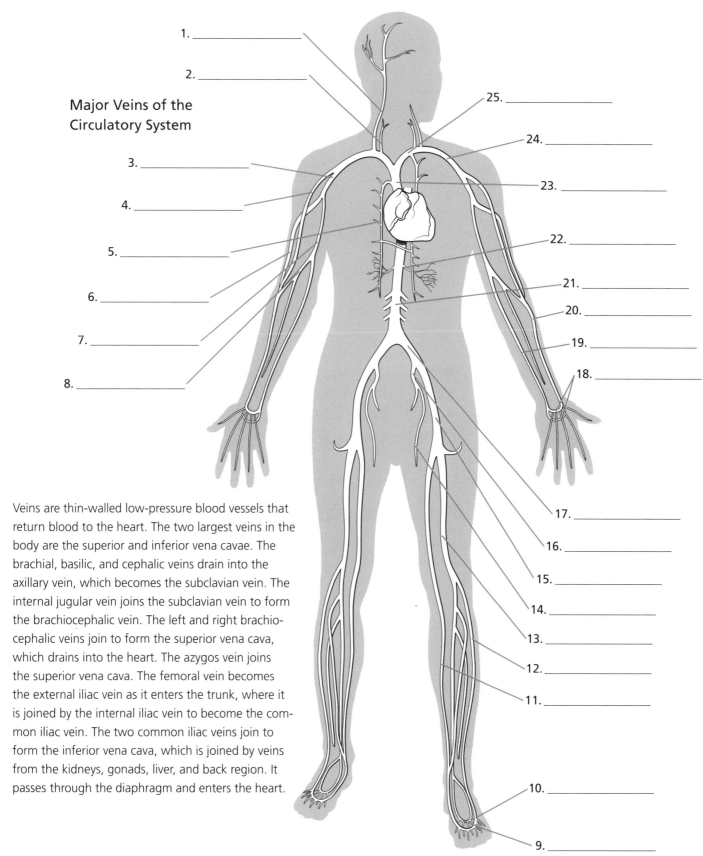

1. _____
2. _____
3. _____
4. _____
5. _____
6. _____
7. _____
8. _____
9. _____
10. _____
11. _____
12. _____
13. _____
14. _____
15. _____
16. _____
17. _____
18. _____
19. _____
20. _____
21. _____
22. _____
23. _____
24. _____
25. _____

Veins are thin-walled low-pressure blood vessels that return blood to the heart. The two largest veins in the body are the superior and inferior vena cavae. The brachial, basilic, and cephalic veins drain into the axillary vein, which becomes the subclavian vein. The internal jugular vein joins the subclavian vein to form the brachiocephalic vein. The left and right brachio-cephalic veins join to form the superior vena cava, which drains into the heart. The azygos vein joins the superior vena cava. The femoral vein becomes the external iliac vein as it enters the trunk, where it is joined by the internal iliac vein to become the com-mon iliac vein. The two common iliac veins join to form the inferior vena cava, which is joined by veins from the kidneys, gonads, liver, and back region. It passes through the diaphragm and enters the heart.

Answers

1. External jugular vein, 2. Internal jugular vein, 3. Axillary vein, 4. Cephalic vein, 5. Azygos vein, 6. Brachial vein, 7. Basilic vein, 8. Median cubital vein, 9. Plantar venous arch, 10. Dorsal venous arch, 11. Great saphenous vein, 12. Small saphenous vein, 13. Femoral vein, 14. Obturator vein, 15. External iliac vein, 16. Internal iliac vein, 17. Common iliac vein, 18. Palmar venous arch, 19. Ulnar vein, 20. Radial vein, 21. Inferior vena cava, 22. Renal vein, 23. Superior vena cava, 24. Subclavian vein, 25. Brachiocephalic vein

Arteries of the Brain

Cerebral Arteries—inferior view

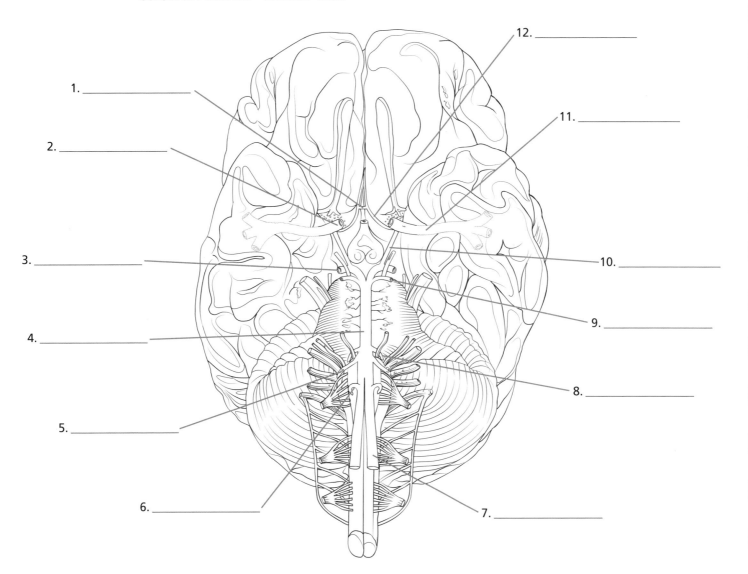

1. _____

2. _____

3. _____

4. _____

5. _____

6. _____

7. _____

8. _____

9. _____

10. _____

11. _____

12. _____

The brain is supplied by a network of arteries known as the cerebral arteries. The cerebral arteries are formed from the carotid and vertebral arteries. The carotid arteries supply the front (anterior cerebral arteries) and middle (middle cerebral arteries) of the brain; the vertebral arteries supply the back of the brain, cerebellum, and brainstem (posterior cerebral and cerebellar arteries). The three paired cerebral arteries (anterior, middle, and posterior) are joined by communicating arteries (anterior and posterior) to form the cerebral arterial circle (of Willis).

Answers

1. Anterior communicating artery, 2. Internal carotid artery, 3. Posterior cerebral artery, 4. Anterior inferior cerebellar artery, 5. Basilar artery, 6. Posterior inferior cerebellar artery, 7. Vertebral artery, 8. Labyrinthine artery, 9. Superior cerebellar artery, 10. Posterior communicating artery, 11. Middle cerebral artery, 12. Anterior cerebral artery

Cerebral Arteries—sagittal view

1. _____

2. _____

3. _____

12. _____

11. _____

10. _____

9. _____

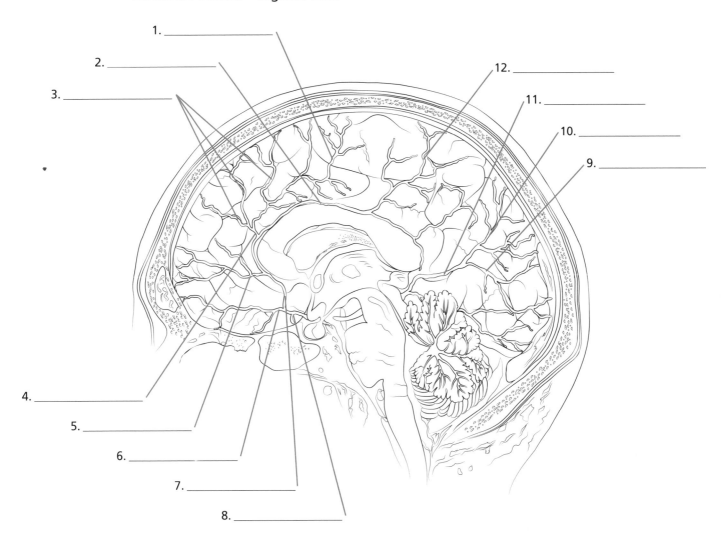

4. _____

5. _____

6. _____

7. _____

8. _____

Each of the three paired arteries establishes a network of branches that supply blood to various parts of the brain and brainstem. Although the brain makes up only 2 percent of the average body weight, it uses 20 percent of the available oxygen.

Answers

Blood Vessels of the Head and Neck

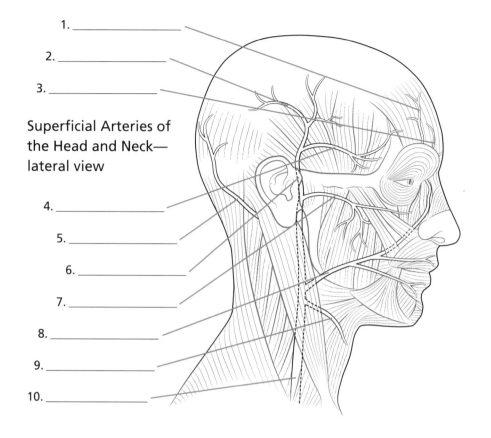

1. _____

2. _____

3. _____

Superficial Arteries of the Head and Neck— lateral view

4. _____

5. _____

6. _____

7. _____

8. _____

9. _____

10. _____

The superficial blood vessels of the head and neck are supplied by branches of the external carotid arteries. Numerous branches lead away from the external carotid artery, supplying blood to the neck, skull, and face. The branches include: the superficial temporal artery, which extends to the top of the head; the facial artery, which extends across the cheek area to the medial side of the eye; the occipital artery, which extends behind the ear; and the maxillary artery, which supplies the jaw, nose, and teeth. The principal superficial veins of the head and neck are the temporal, facial, and occipital veins. These veins drain the superficial area above, in front of, and behind the ear, respectively; the area around the cheeks and nose; and the area around the jaw. These veins drain into the external jugular veins.

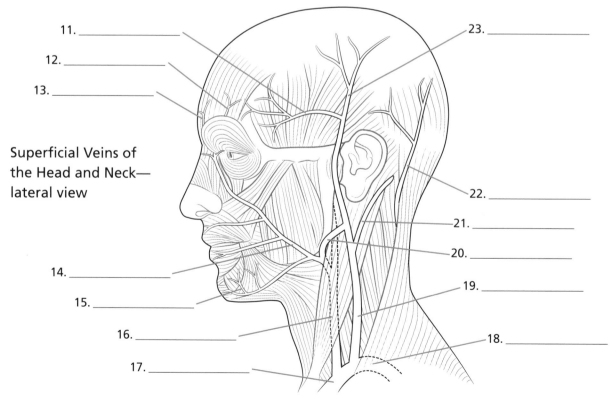

11. _____

12. _____

13. _____

Superficial Veins of the Head and Neck— lateral view

14. _____

15. _____

16. _____

17. _____

23. _____

22. _____

21. _____

20. _____

19. _____

18. _____

Answers

1. Supraorbital artery, 2. Posterior branch of superficial temporal artery, 3. Supratrochlear artery, 4. Anterior branch of superficial temporal artery, 5. Occipital artery, 6. Superficial temporal artery, 7. Transverse cervical artery, 8. Facial artery, 9. Facial artery, 10. External carotid artery, 11. Anterior branch of superficial temporal vein, 12. Supraorbital vein, 13. Supratrochlear vein, 14. Facial vein, 15. Submental vein, 16. Internal jugular vein, 17. Brachiocephalic vein, 18. Subclavian vein, 19. External jugular vein, 20. Retromandibular vein, 21. Posterior auricular vein, 22. Occipital vein, 23. Posterior branch of superficial temporal vein

**Blood Vessels of the Eye—
lateral view**

8. _____

7. _____

6. _____

5. _____

1. _____

2. _____

3. _____

4. _____

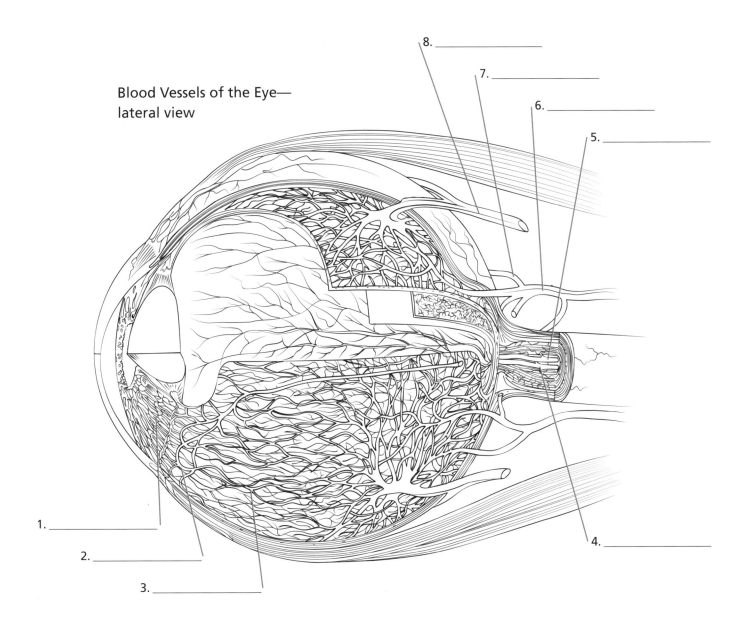

The central artery of the retina enters the eyeball after running through the center of the optic nerve. As it emerges from the optic disk, the central artery of the retina fans out into four main branches accompanied by their veins. These branches run on the inside of the retina and spread out into the capillary network. The central vein of the retina receives blood from the retinal veins. Blood from the central vein of the retina drains into the superior ophthalmic vein.

Answers

Heart

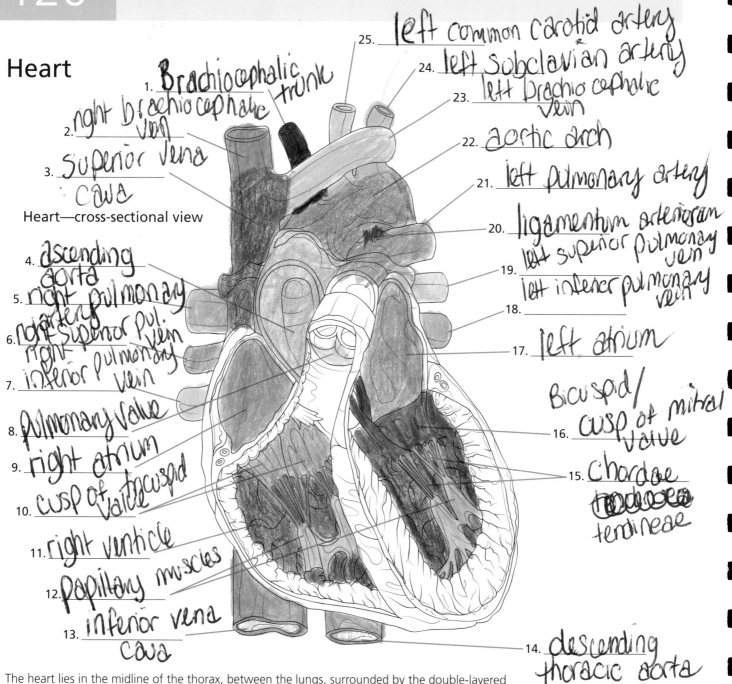

Heart—cross-sectional view

1. Brachiocephalic trunk
2. right brachiocephalic vein
3. superior vena cava
4. ascending aorta
5. right pulmonary artery
6. right superior pul. vein
7. right inferior pulmonary vein
8. Pulmonary valve
9. right atrium
10. cusp of valve/cusp of tricuspid
11. right ventricle
12. Papillary muscles
13. inferior vena cava
14. descending thoracic aorta

15. Chordae tendineae
16. Bicuspid/cusp of mitral valve
17. left atrium
18. _____
19. left inferior pulmonary vein
20. ligamentum arteriosum / left superior pulmonary vein
21. left pulmonary artery
22. aortic arch
23. left brachiocephalic vein
24. left subclavian artery
25. left common carotid artery

The heart lies in the midline of the thorax, between the lungs, surrounded by the double-layered membrane of the pericardium. The right and left atria are located at the back, and the right and left ventricles at the front. A septum divides the right and left atria and the right and left ventricles. Each atrium opens into a ventricle through an atrioventricular orifice, which is guarded by a valve to ensure that blood flows in only one direction. The left atrium and ventricle are responsible for receiving oxygen-rich blood from the lungs and pumping it out to the body through the aorta. The right atrium and ventricle are responsible for receiving relatively deoxygenated blood from the body and distributing it via the pulmonary trunk to the lungs for gas exchange to occur. A cross-sectional view of the heart exposes the four chambers and valves. The mitral valve sits between the left atrium and ventricle. Blood exiting from the left ventricle is released through the aortic valve. The tricuspid valve sits between the right atrium and ventricle. Blood leaving the right ventricle passes through the pulmonary valve.

Answers

1. Brachiocephalic artery, 2. Right brachiocephalic vein, 3. Superior vena cava, 4. Ascending aorta, 5. Right pulmonary artery, 6. Right superior pulmonary vein, 7. Right inferior pulmonary vein, 8. Pulmonary valve, 9. Right atrium, 10. Cusp of tricuspid valve, 11. Right ventricle, 12. Papillary muscles, 13. Inferior vena cava, 14. Descending thoracic aorta, 15. Chordae tendineae, 16. Cusp of mitral valve, 17. Left atrium, 18. Left atrium, 19. Left inferior pulmonary vein, 20. Ligamentum arteriosum, 21. Left pulmonary artery, 22. Aortic arch, 23. Left brachiocephalic vein, 24. Left subclavian artery, 25. Left common carotid artery

1. brachiocephalic artery
2. right brachiocephalic vein
3. superior vena cava
4. right atrium
5. right pulmonary artery
6. right superior pulmonary vein
7. right inferior pulmonary vein
8. right coronary artery
9. right ~~left~~ ventricle
10. inferior vena cava
11. descending aorta

21. left common carotid artery
20. left subclavian artery
19. left brachiocephalic vein
18. aortic arch
17. left pulmonary artery
16. left superior pulmonary vein
15. left inferior pulmonary vein
14. ~~left~~ left atrium
13. branch of left coronary artery
12. left ventricle

Heart—anterior view

22. left subclavian artery
23. aortic arch
24. left pulmonary artery
25. pericardium
26. right pulmonary artery
27. left superior pulmonary vein
28. left inferior pulmonary vein
29. left ventricle

38. left common carotid artery
37. brachiocephalic artery
36. superior vena cava
35. right superior pulmonary vein
34. right inferior pulmonary vein
33. right atrium
32. inferior vena cava
31. right coronary artery
30. right ventricle

Heart—posterior view

Heart

Heart Valves: Ventricular Systole

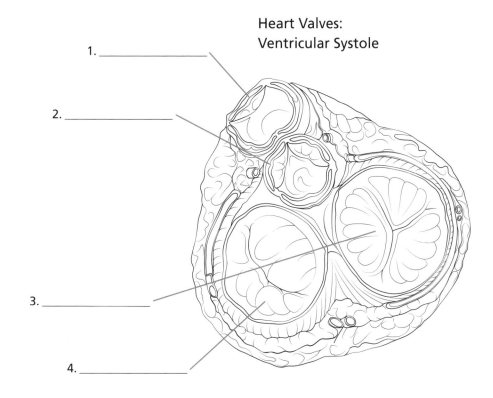

1. _____
2. _____
3. _____
4. _____

The four valves of the heart are designed to allow one-way flow of blood. The atrioventricular valves—the mitral and tricuspid valves—separate the atrium and ventricle on the left and right sides of the heart, respectively. During ventricular systole, the ventricles of the heart contract and the pulmonary and aortic valves open to allow blood to be pumped into the pulmonary and general circulatory systems, respectively, while the mitral and tricuspid valves remain closed. During ventricular diastole, the aortic and pulmonary valves close, while the atrioventricular valves (the tricuspid and mitral valves) open to allow blood to pass from the atria to the ventricles. The cusps of the atrioventricular valves have tough fibrous cords, the chordae tendineae, which are attached along the free edge of a valve. The other ends of the chordae tendineae are attached to the papillary muscles.

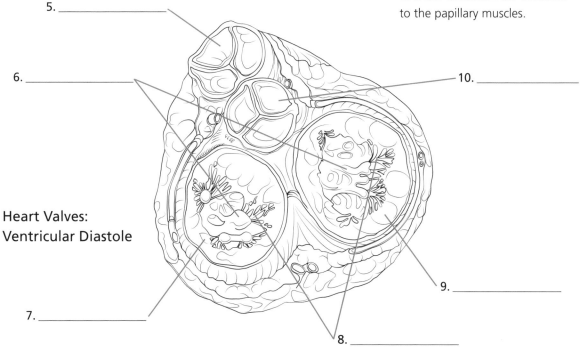

5. _____
6. _____
7. _____
8. _____
9. _____
10. _____

Heart Valves: Ventricular Diastole

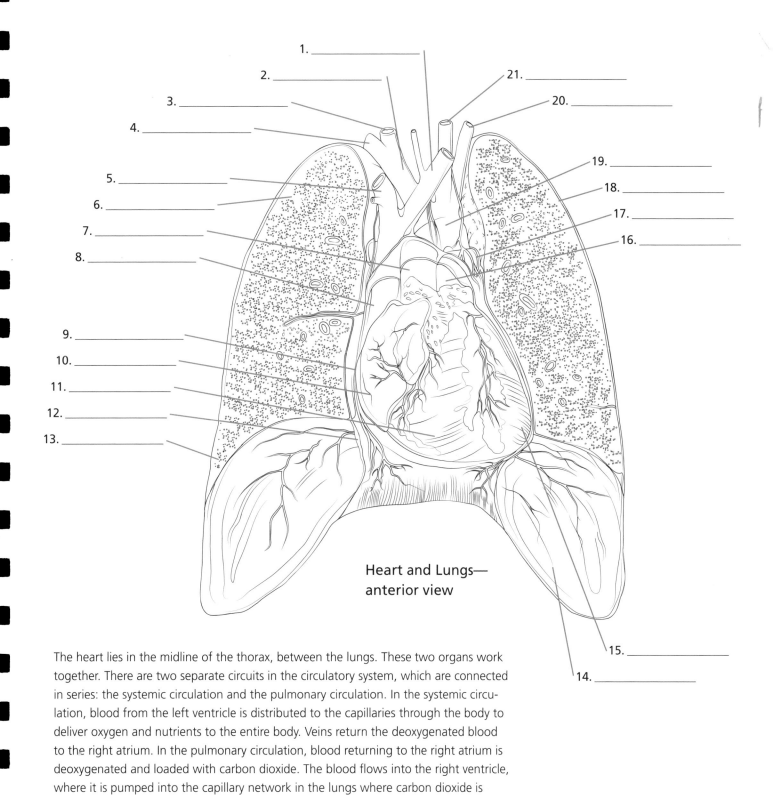

Heart and Lungs—
anterior view

1. _____
2. _____
3. _____
4. _____
5. _____
6. _____
7. _____
8. _____
9. _____
10. _____
11. _____
12. _____
13. _____
14. _____
15. _____
16. _____
17. _____
18. _____
19. _____
20. _____
21. _____

The heart lies in the midline of the thorax, between the lungs. These two organs work together. There are two separate circuits in the circulatory system, which are connected in series: the systemic circulation and the pulmonary circulation. In the systemic circulation, blood from the left ventricle is distributed to the capillaries through the body to deliver oxygen and nutrients to the entire body. Veins return the deoxygenated blood to the right atrium. In the pulmonary circulation, blood returning to the right atrium is deoxygenated and loaded with carbon dioxide. The blood flows into the right ventricle, where it is pumped into the capillary network in the lungs where carbon dioxide is exchanged with oxygen. Oxygenated blood returns to the left atrium.

Answers

1. Left brachiocephalic vein, 2. Brachiocephalic trunk, 3. Right common carotid artery, 4. Right subclavian artery, 5. Right brachiocephalic vein, 6. Right lung (upper lobe), 7. Ascending aorta, 8. Superior vena cava, 9. Pericardium, 10. Right atrium, 11. Right ventricle, 12. Pleura, 13. Right lung (lower lobe), 14. Diaphragm, 15. Left ventricle, 16. Pulmonary trunk, 17. Left pulmonary artery, 18. Left lung (upper lobe), 19. Aortic arch, 20. Left subclavian artery, 21. Left common carotid artery

Blood Vessels of the Abdominal Region

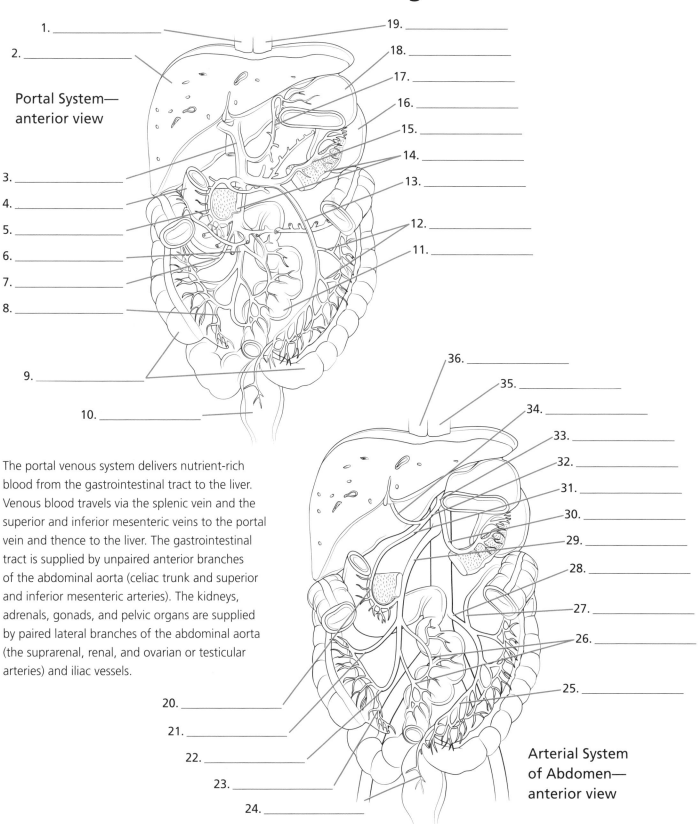

Portal System—
anterior view

1. _____
2. _____
3. _____
4. _____
5. _____
6. _____
7. _____
8. _____
9. _____
10. _____

19. _____
18. _____
17. _____
16. _____
15. _____
14. _____
13. _____
12. _____
11. _____

The portal venous system delivers nutrient-rich blood from the gastrointestinal tract to the liver. Venous blood travels via the splenic vein and the superior and inferior mesenteric veins to the portal vein and thence to the liver. The gastrointestinal tract is supplied by unpaired anterior branches of the abdominal aorta (celiac trunk and superior and inferior mesenteric arteries). The kidneys, adrenals, gonads, and pelvic organs are supplied by paired lateral branches of the abdominal aorta (the suprarenal, renal, and ovarian or testicular arteries) and iliac vessels.

20. _____
21. _____
22. _____
23. _____
24. _____

36. _____
35. _____
34. _____
33. _____
32. _____
31. _____
30. _____
29. _____
28. _____
27. _____
26. _____
25. _____

Arterial System
of Abdomen—
anterior view

Answers

1. Inferior vena cava, 2. Liver, 3. Portal vein, 4. Duodenum, 5. Pancreaticoduodenal vein, 6. Superior mesenteric vein, 7. Right colic vein, 8. Appendicular vein, 9. Colon, 10. Rectum, 11. Small intestine, 12. Left colic veins, 13. Inferior mesenteric vein, 14. Pancreas, 15. Spleen, 16. Splenic vein, 17. Left gastric vein, 18. Stomach, 19. Thoracic aorta, 20. Pancreaticoduodenal artery, 21. Right colic artery, 22. Iliocolic artery, 23. Appendicular artery, 24. Rectal artery, 25. Sigmoidal artery, 26. Jejunal arteries, 27. Left colic artery, 28. Inferior mesenteric artery, 29. Superior mesenteric artery, 30. Splenic artery, 31. Gastroduodenal artery, 32. Common hepatic artery, 33. Celiac trunk, 34. Hepatic artery proper, 35. Thoracic aorta, 36. Inferior vena cava

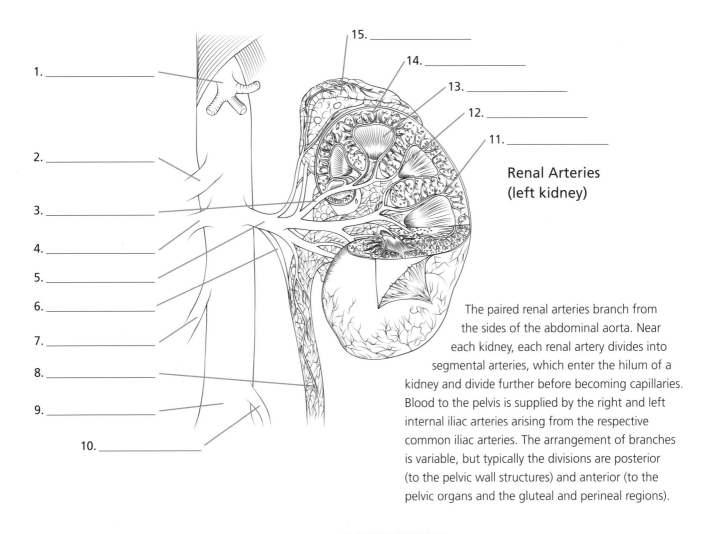

1. _____

2. _____

3. _____

4. _____

5. _____

6. _____

7. _____

8. _____

9. _____

10. _____

15. _____

14. _____

13. _____

12. _____

11. _____

Renal Arteries (left kidney)

The paired renal arteries branch from the sides of the abdominal aorta. Near each kidney, each renal artery divides into segmental arteries, which enter the hilum of a kidney and divide further before becoming capillaries. Blood to the pelvis is supplied by the right and left internal iliac arteries arising from the respective common iliac arteries. The arrangement of branches is variable, but typically the divisions are posterior (to the pelvic wall structures) and anterior (to the pelvic organs and the gluteal and perineal regions).

Arteries of the Pelvic Wall

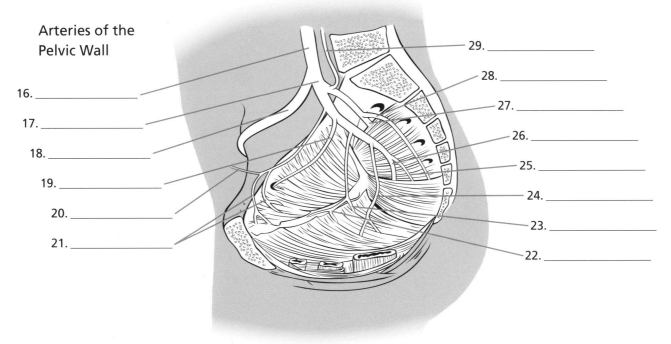

16. _____

17. _____

18. _____

19. _____

20. _____

21. _____

29. _____

28. _____

27. _____

26. _____

25. _____

24. _____

23. _____

22. _____

Answers

1. Celiac trunk, 2. Superior mesenteric artery, 3. Segmental artery, 4. Right gonadal artery, 5. Left renal artery, 6. Left gonadal artery, 7. Right gonadal artery, 8. Ureter, 9. Abdominal aorta, 10. Inferior mesenteric artery, 11. Arcuate artery, 12. Interlobar artery, 13. Renal pyramid (medulla), 14. Cortex, 15. Left adrenal gland, 16. Common iliac artery, 17. Internal iliac artery, 18. External iliac artery, 19. Obturator artery, 20. Obliterated umbilical artery, 21. Superior vesicle, 22. Uterine artery, 23. Vaginal artery, 24. Middle rectal artery, 25. Internal pudendal artery, 26. Inferior gluteal artery, 27. Superior gluteal artery, 28. Lateral sacral artery, 29. Iliolumbar artery

Blood Vessels of the Upper and Lower Limbs

23. subclavian vein
22. axillary vein

Veins of the Upper Limb—anterior view

1. _____

2. _____

3. _____

15. _____

16. brachial vein

17. _____

4. _____

5. _____

6. _____

18. _____

19. _____

Arteries of the Upper Limb—anterior view

7. _____

8. _____

9. _____

20. _____

10. _____

11. _____

12. _____

13. _____

21. _____

14. _____

The brachial artery is the major artery of the upper limb. It courses down the length of the arm, dividing at the elbow into two major branches—the radial artery and the ulnar artery—that supply the forearm and hand. In the upper limb, the veins are organized into two groups: the deep group and the superficial group. The deep group veins travel with the deep arteries and are similarly named. The superficial group veins course immediately beneath the skin. Blood from the digital veins drains into the palmar venous arch, which in turn drains into the cephalic and basilic veins. The basilic vein joins up with the brachial vein and, along with the cephalic vein, drains into the axillary vein.

Answers

1. Subclavian artery, 2. Lateral thoracic artery, 3. Subscapular artery, 4. Axillary artery, 5. Deep brachial artery, 6. Brachial artery, 7. Superior ulnar collateral artery, 8. Inferior ulnar collateral artery, 9. Interosseous artery, 10. Ulnar artery, 11. Radial artery, 12. Superficial palmar arch, 13. Deep palmar arch, 14. Digital arteries, 15. Cephalic vein, 16. Brachial vein, 17. Basilic vein, 18. Median cubital vein, 19. Median antebrachial vein, 20. Palmar venous arch, 21. Digital veins, 22. Axillary vein, 23. Subclavian vein

Veins of the Lower Limb—
anterior view

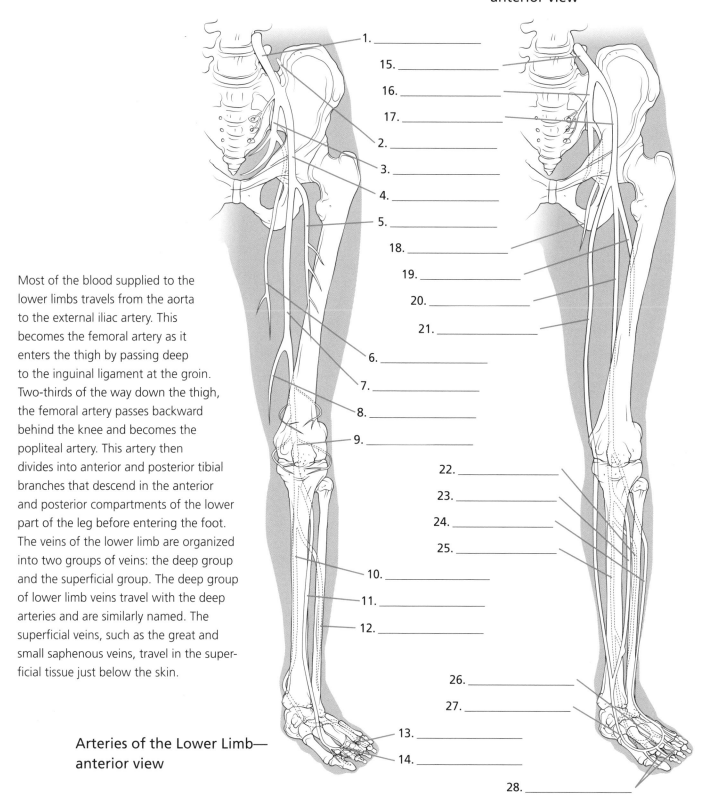

1. _____

15. _____

16. _____

17. _____

2. _____

3. _____

4. _____

5. _____

18. _____

19. _____

20. _____

21. _____

6. _____

7. _____

8. _____

9. _____

22. _____

23. _____

24. _____

25. _____

10. _____

11. _____

12. _____

26. _____

27. _____

13. _____

14. _____

28. _____

Most of the blood supplied to the lower limbs travels from the aorta to the external iliac artery. This becomes the femoral artery as it enters the thigh by passing deep to the inguinal ligament at the groin. Two-thirds of the way down the thigh, the femoral artery passes backward behind the knee and becomes the popliteal artery. This artery then divides into anterior and posterior tibial branches that descend in the anterior and posterior compartments of the lower part of the leg before entering the foot. The veins of the lower limb are organized into two groups of veins: the deep group and the superficial group. The deep group of lower limb veins travel with the deep arteries and are similarly named. The superficial veins, such as the great and small saphenous veins, travel in the super-ficial tissue just below the skin.

Arteries of the Lower Limb—
anterior view

Answers

1. Common iliac artery, 2. Iliolumbar artery, 3. Internal iliac artery, 4. External iliac artery, 5. Deep femoral artery, 6. Obturator artery, 7. Femoral artery, 8. Descending genicular artery, 9. Popliteal artery, 10. Posterior tibial artery, 11. Anterior tibial artery, 12. Fibular artery, 13. Plantar arch, 14. Arcuate artery, 15. Common iliac vein, 16. Internal iliac vein, 17. External iliac vein, 18. Obturator vein, 19. Profunda femoris vein, 20. Femoral vein, 21. Great saphenous vein, 22. Fibular vein, 23. Anterior tibial vein, 24. Small saphenous vein, 25. Posterior tibial vein, 26. Plantar venous arch, 27. Dorsal venous arch, 28. Digital veins

Respiratory System

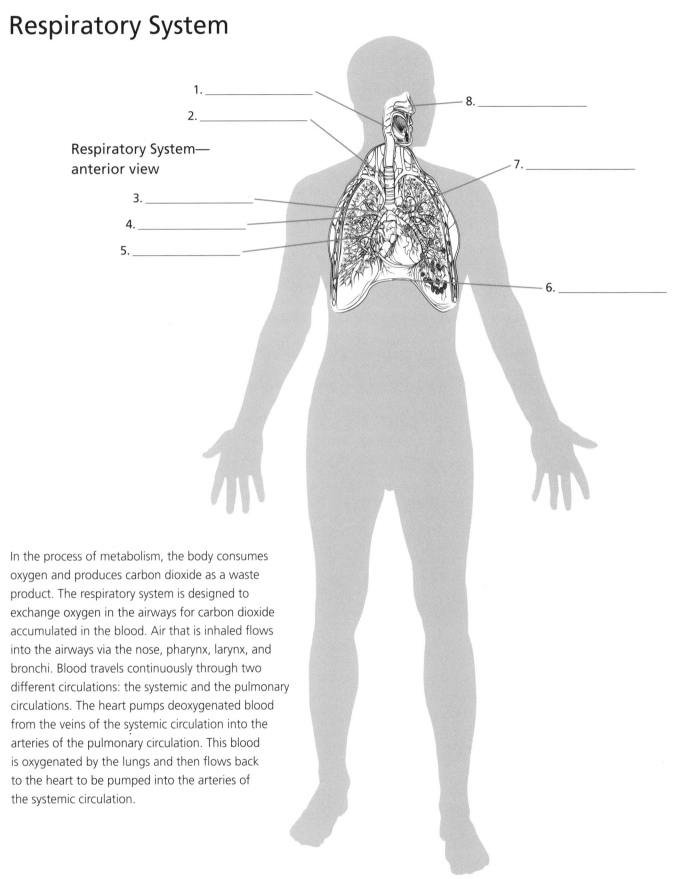

Respiratory System—
anterior view

1. _____

2. _____

3. _____

4. _____

5. _____

6. _____

7. _____

8. _____

In the process of metabolism, the body consumes oxygen and produces carbon dioxide as a waste product. The respiratory system is designed to exchange oxygen in the airways for carbon dioxide accumulated in the blood. Air that is inhaled flows into the airways via the nose, pharynx, larynx, and bronchi. Blood travels continuously through two different circulations: the systemic and the pulmonary circulations. The heart pumps deoxygenated blood from the veins of the systemic circulation into the arteries of the pulmonary circulation. This blood is oxygenated by the lungs and then flows back to the heart to be pumped into the arteries of the systemic circulation.

Answers

1. Pharynx, 2. Trachea, 3. Right primary bronchus, 4. Superior lobar bronchus, 5. Middle lobar bronchus, 6. Diaphragm, 7. Left primary bronchus, 8. Nasal cavity

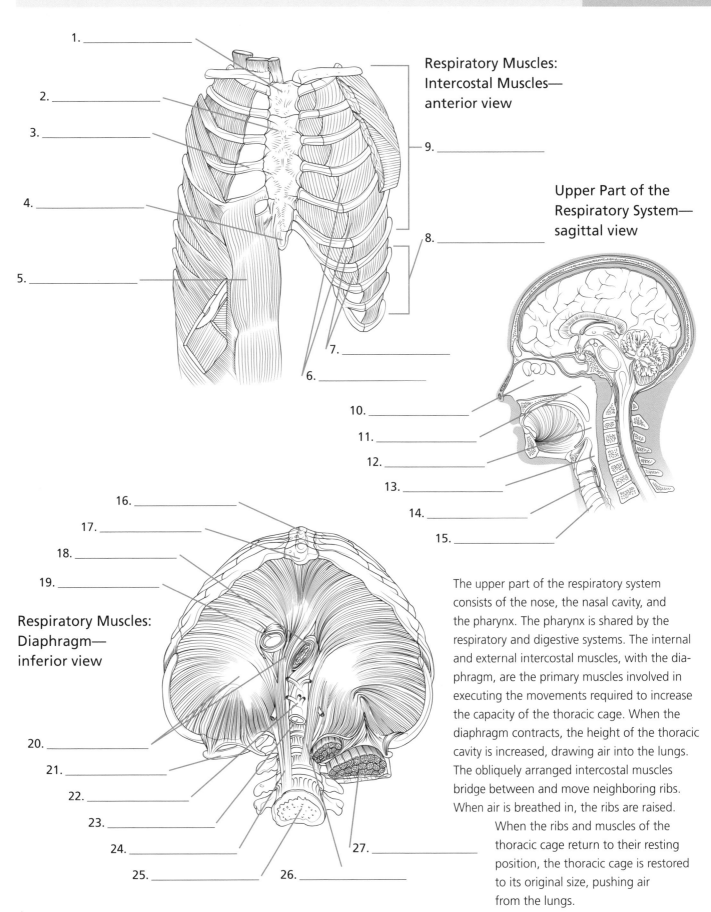

Respiratory Muscles:
Intercostal Muscles—
anterior view

Upper Part of the
Respiratory System—
sagittal view

1. _____
2. _____
3. _____
4. _____
5. _____
6. _____
7. _____
8. _____
9. _____
10. _____
11. _____
12. _____
13. _____
14. _____
15. _____

Respiratory Muscles:
Diaphragm—
inferior view

16. _____
17. _____
18. _____
19. _____
20. _____
21. _____
22. _____
23. _____
24. _____
25. _____
26. _____
27. _____

The upper part of the respiratory system
consists of the nose, the nasal cavity, and
the pharynx. The pharynx is shared by the
respiratory and digestive systems. The internal
and external intercostal muscles, with the diaphragm, are the primary muscles involved in
executing the movements required to increase
the capacity of the thoracic cage. When the
diaphragm contracts, the height of the thoracic
cavity is increased, drawing air into the lungs.
The obliquely arranged intercostal muscles
bridge between and move neighboring ribs.
When air is breathed in, the ribs are raised.

When the ribs and muscles of the
thoracic cage return to their resting
position, the thoracic cage is restored
to its original size, pushing air
from the lungs.

Answers

1. Manubrium, 2. Body of sternum, 3. Costal cartilage, 4. Xiphoid process, 5. Rectus abdominus muscle, 6. External intercostal muscles, 7. Internal intercostal muscles, 8. Ribs 8–10 ("floating" ribs), 9. Ribs 1–7 ("true" ribs), 10. Nasal cavity, 11. Nasopharynx, 12. Oropharynx, 13. Laryngopharynx, 14. Larynx, 15. Trachea, 16. Body of sternum, 17. Xiphoid process, 18. Esophagus, 19. Inferior vena cava, 20. Central tendon, 21. Twelfth rib, 22. Celiac trunk, 23. Abdominal aorta, 24. Right crus of diaphragm, 25. Vertebral column, 26. Left crus of diaphragm, 27. Quadratus lumborum muscle

Nose, Pharynx, and Larynx

Lined with ciliated mucous membrane, the paranasal sinuses are air-containing cavities within the frontal, sphenoid, ethmoid, and maxillary bones of the skull, which are connected by passages to the nose. The paranasal sinuses are rudimentary at birth and develop rapidly at the age of puberty. The lacrimal glands, one above each eye, produce fluid (tears), which moves across the eye from the outer corner toward the nose, lubricating the eyeball. The tears eventually drain into the nose through the nasolacrimal duct and are swallowed or blown out with other nasal fluids.

1. _____

2. _____

3. _____

4. _____

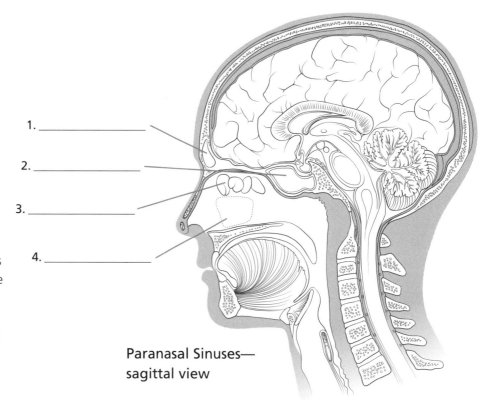

**Paranasal Sinuses—
sagittal view**

5. _____

6. _____

7. _____

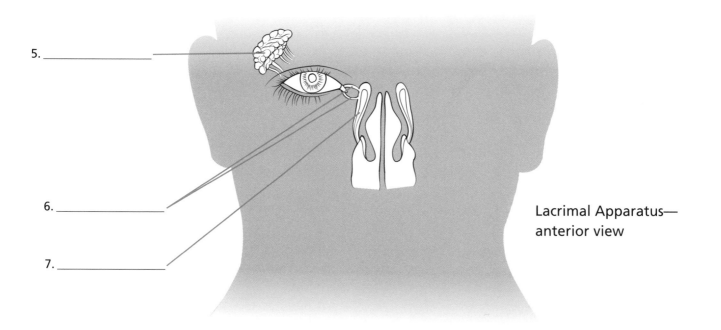

**Lacrimal Apparatus—
anterior view**

Answers

1. Frontal sinus, 2. Sphenoid sinus, 3. Ethmoid sinuses, 4. Maxillary sinus, 5. Lacrimal gland, 6. Lacrimal canals, 7. Nasolacrimal duct

Pharynx—posterior view

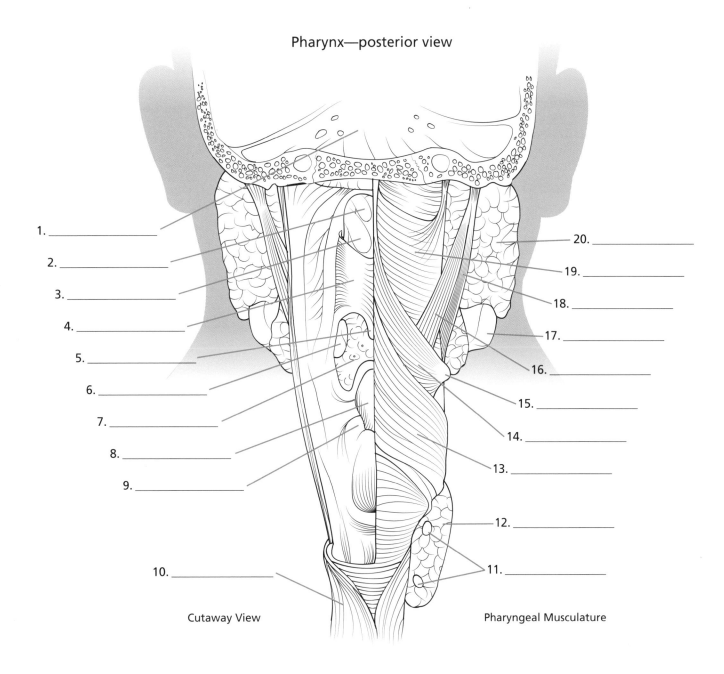

1. _____
2. _____
3. _____
4. _____
5. _____
6. _____
7. _____
8. _____
9. _____
10. _____

Cutaway View

20. _____
19. _____
18. _____
17. _____
16. _____
15. _____
14. _____
13. _____
12. _____
11. _____

Pharyngeal Musculature

The pharynx lies behind the nasal cavity, oral cavity, and larynx and ends in the esophagus. The pharynx has openings to all these regions and is a common passage for air, liquid, and food. Located on each side wall of the pharynx, just behind the nasal cavity, is the opening of the auditory tube, which connects the pharynx with the middle ear cavity. The pharynx has three parts: the nasopharynx, oropharynx, and laryngo-pharynx. The nasopharynx lies immediately beneath the base of the skull and behind the nose. The oropharynx lies behind the mouth. The laryngopharynx, or hypopharynx, is the lowest part of the pharynx and extends from the tip of the epiglottic cartilage to the lower edge of the larynx.

Answers

1. Base of skull, 2. Middle nasal concha, 3. Inferior nasal concha, 4. Soft palate, 5. Uvula, 6. Palatine tonsil, 7. Dorsum of tongue, 8. Epiglottis, 9. Aryepiglottic fold, 10. Esophagus, 11. Parathyroid glands, 12. Thyroid gland (lateral lobe), 13. Inferior constrictor muscle, 14. Middle constrictor muscle, 15. End of greater horn of hyoid bone, 16. Stylopharyngeus muscle, 17. Angle of mandible, 18. Stylohyoid muscle, 19. Superior constrictor muscle, 20. Parotid gland

Nose, Pharynx, and Larynx

The larynx extends from the back of the tongue to the trachea. The larynx is composed of nine cartilages that provide strength for the airway and attachments for the muscles, ligaments, and membranes of the larynx. The upper segment of the larynx commences at the epiglottis and ends at the vestibular folds or "false vocal folds" (false vocal cords). The lower segment begins at the true vocal folds (true vocal cords) and ends at the cricoid cartilage. The cricoid cartilage, attached to the trachea below, is the only complete ring of cartilage in the respiratory system. The vocal cords are attached to the thyroid cartilage at the front and to the arytenoid cartilages at the back. The thyroid cartilage, which is connected by a membrane to the hyoid bone above and to the cricoid cartilage below, protects the front of the larynx.

Larynx—sagittal view

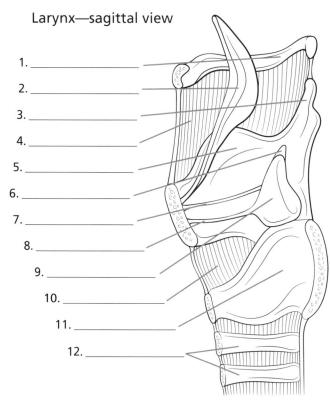

1. _____
2. _____
3. _____
4. _____
5. _____
6. _____
7. _____
8. _____
9. _____
10. _____
11. _____
12. _____

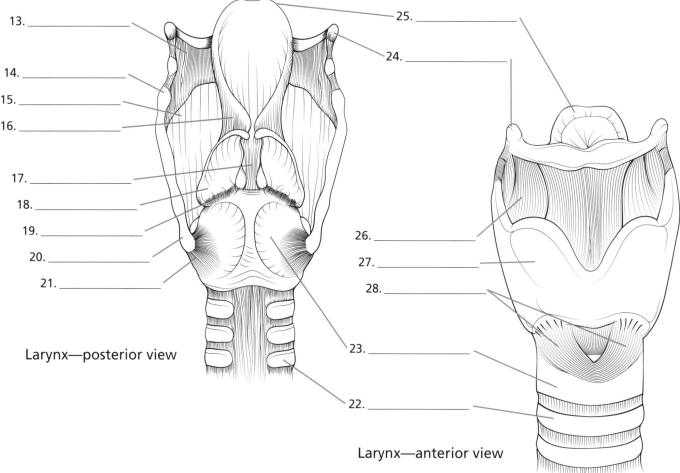

13. _____
14. _____
15. _____
16. _____
17. _____
18. _____
19. _____
20. _____
21. _____

Larynx—posterior view

25. _____
24. _____

26. _____
27. _____
28. _____
23. _____
22. _____

Larynx—anterior view

Answers

1. Hyoid bone, 2. Epiglottis, 3. Superior horn of thyroid cartilage, 4. Thyrohyoid membrane, 5. Lamina of thyroid cartilage, 6. Corniculate cartilage, 7. False vocal fold, 8. True vocal fold, 9. Arytenoid cartilage, 10. Cricothyroid membrane, 11. Cricoid cartilage, 12. Tracheal cartilages, 13. Superior horn of thyroid cartilage, 14. Thyrohyoid membrane, 15. Lamina of thyroid cartilage, 16. Quadrangular membrane, 17. Stem of epiglottis, 18. Arytenoid cartilage, 19. Capsule of cricoarytenoid joint, 20. Inferior horn of thyroid cartilage, 21. Capsule of cricothyroid joint, 22. Tracheal cartilage, 23. Cricoid cartilage, 24. Greater horn of hyoid bone, 25. Epiglottis, 26. Thyrohyoid membrane, 27. Thyroid cartilage, 28. Cricothyroid muscle

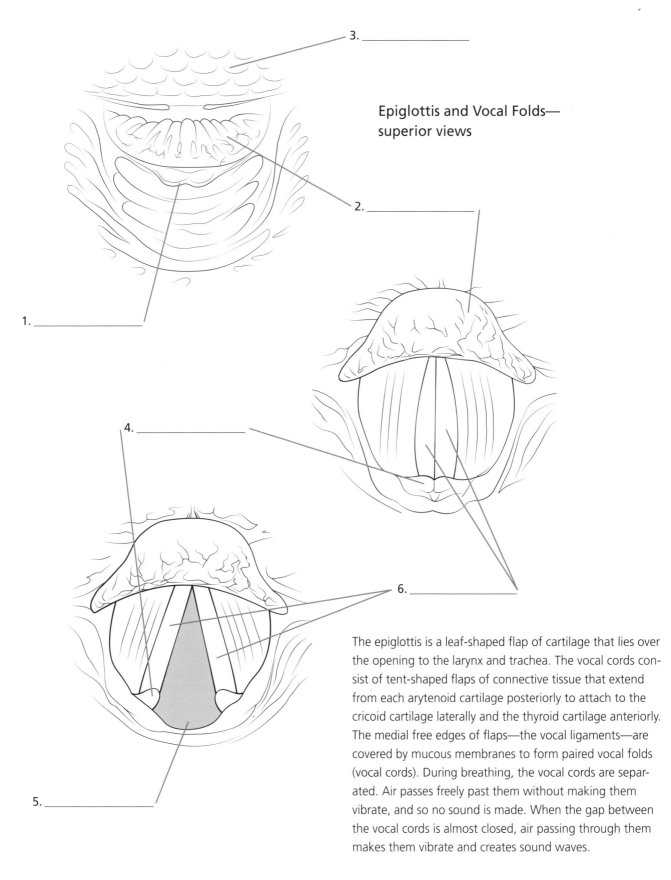

3. _____

**Epiglottis and Vocal Folds—
superior views**

2. _____

1. _____

4. _____

6. _____

5. _____

The epiglottis is a leaf-shaped flap of cartilage that lies over the opening to the larynx and trachea. The vocal cords consist of tent-shaped flaps of connective tissue that extend from each arytenoid cartilage posteriorly to attach to the cricoid cartilage laterally and the thyroid cartilage anteriorly. The medial free edges of flaps—the vocal ligaments—are covered by mucous membranes to form paired vocal folds (vocal cords). During breathing, the vocal cords are separated. Air passes freely past them without making them vibrate, and so no sound is made. When the gap between the vocal cords is almost closed, air passing through them makes them vibrate and creates sound waves.

Answers

1. Corniculate tubercle, 2. Epiglottis, 3. Root of tongue, 4. Vocal process of arytenoid cartilage, 5. Trachea, 6. Vocal folds

Trachea and Bronchi

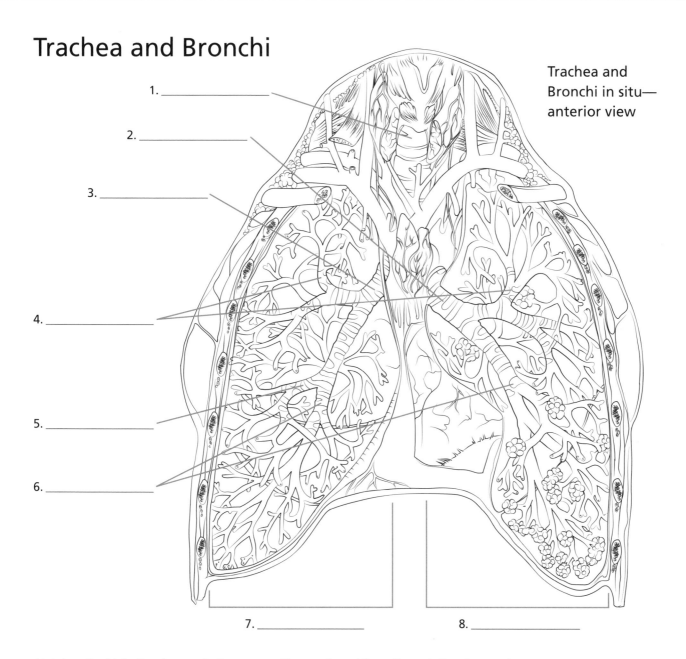

1. _____

2. _____

3. _____

4. _____

5. _____

6. _____

7. _____

8. _____

Trachea and Bronchi in situ— anterior view

Air is breathed into the airways via the nose and/or mouth and flows through the pharynx, larynx, trachea, and bronchi. At its lower end, the trachea branches into the right primary bronchus and the left primary bronchus. Each primary bronchus branches into smaller secondary bronchi at the lung's entrance. Each secondary bronchus divides into smaller tertiary bronchi. These bronchi split into subsequent generations of smaller and smaller bronchi, which finally split into bronchioles that separate into terminal bronchioles. The bronchioles end at the microscopic air sacs (alveoli) where the exchange of carbon dioxide for oxygen occurs. The trachea is reinforced at the front and sides by C-shaped plates of cartilage. Bridging the ends of the cartilage is the trachealis muscle. The internal lining of the trachea is respiratory epithelium. Similar in composition to the trachea, the bronchi and bronchioles feature a muscular exterior lined with respiratory epithelium. Cartilage also forms part of the structure of bronchi, but as the bronchi subdivide the amount of cartilage gradually lessens, and when the bronchioles are reached there is no cartilage present.

Answers

1. Trachea, 2. Left primary bronchi, 3. Right primary bronchi, 4. Superior lobar bronchus, 5. Middle lobar bronchi, 6. Inferior lobar bronchus, 7. Right lung, 8. Left lung

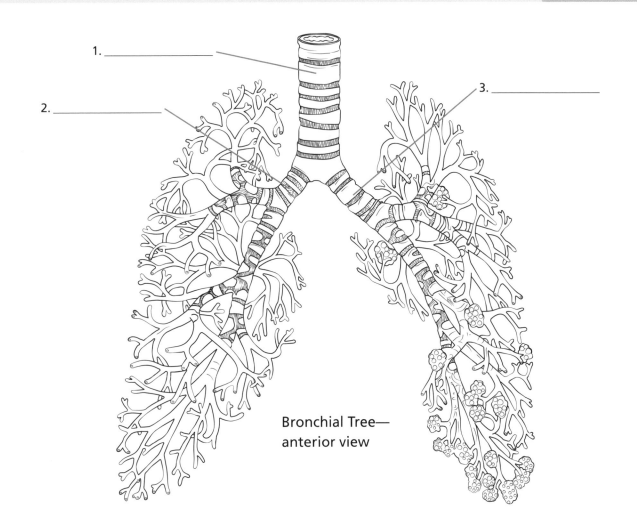

Bronchial Tree—
anterior view

1. _____

2. _____

3. _____

Trachea—cross-sectional view

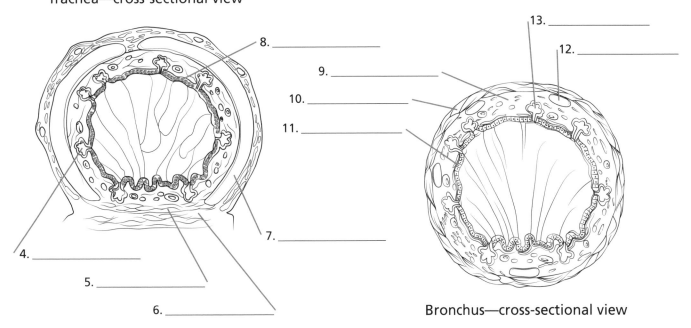

8. _____

9. _____

10. _____

11. _____

7. _____

4. _____

5. _____

6. _____

13. _____

12. _____

Bronchus—cross-sectional view

Lungs

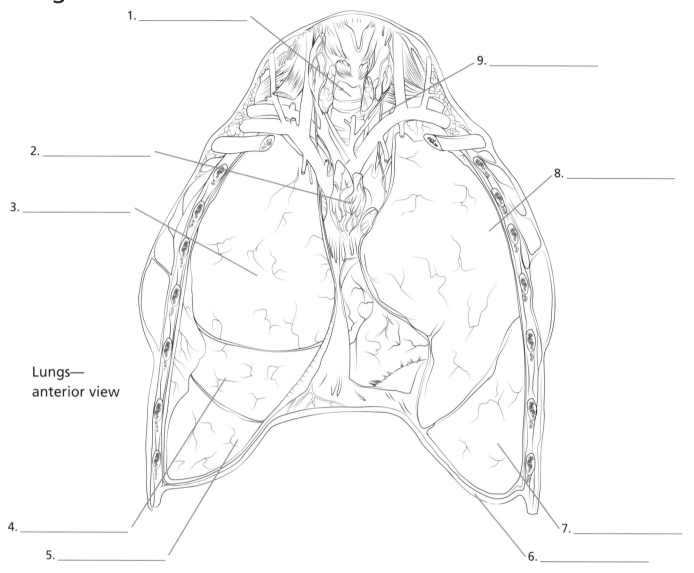

1. _____

9. _____

2. _____

8. _____

3. _____

Lungs—
anterior view

4. _____

5. _____

7. _____

6. _____

The lungs are the two main organs of the respiratory system, lying on either side of the heart within the chest cavity. The lungs are enclosed within pleural sacs. Each lung has a roughly conical or pyramidal shape, with a base sitting on top of the diaphragm; sides in contact with the rib cage (costal surface), the mediastinum (mediastinal surface), and the backbone (vertebral surface); and an apex. The lung apex is encircled by the first rib and lies above the first rib in the hollow at the angle between the neck and the shoulder. On the mediastinal surface of each lung is the lung hilum, where the left and right main bronchi enter the lung and the pulmonary arteries and veins enter and exit, respectively. The lungs are divided into lobes—usually three in the right lung and two in the left lung—by a series of clefts or fissures. The mediastinum contains the heart and pericardium, esophagus, trachea, and associated blood vessels and nerves. The mediastinal organs are encircled by the sternum of the thoracic cage at the front, by the lungs and pleura (pleural cavities) on each side, and by the vertebral column at the back.

Answers

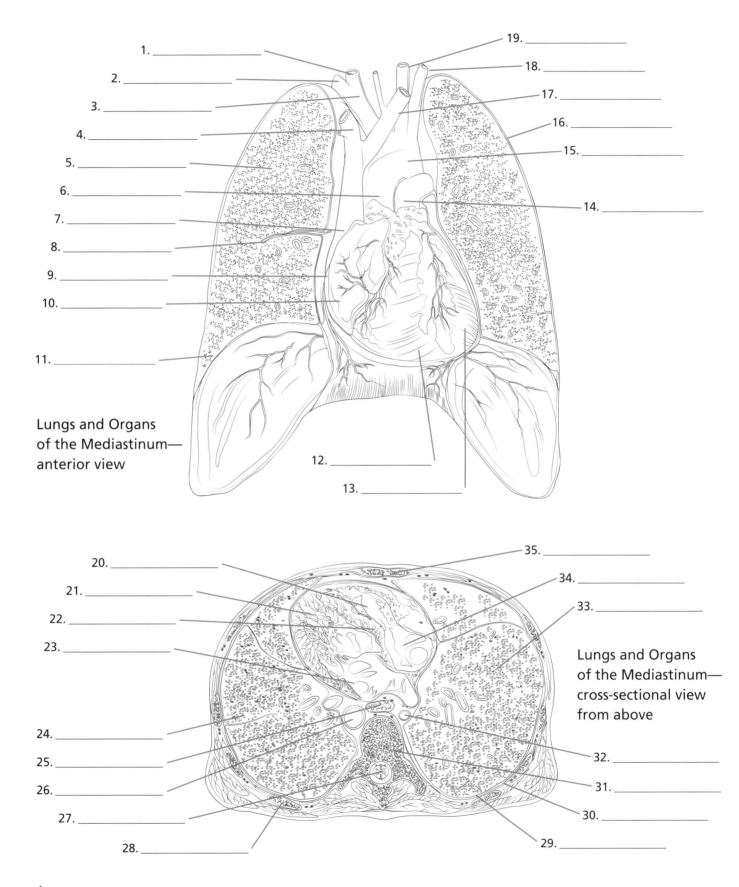

1. _____

2. _____

3. _____

4. _____

5. _____

6. _____

7. _____

8. _____

9. _____

10. _____

11. _____

12. _____

13. _____

14. _____

15. _____

16. _____

17. _____

18. _____

19. _____

Lungs and Organs
of the Mediastinum—
anterior view

20. _____

21. _____

22. _____

23. _____

24. _____

25. _____

26. _____

27. _____

28. _____

29. _____

30. _____

31. _____

32. _____

33. _____

34. _____

35. _____

Lungs and Organs
of the Mediastinum—
cross-sectional view
from above

Answers

1. Right common carotid artery, 2. Right subclavian artery, 3. Brachiocephalic trunk, 4. Right brachiocephalic vein, 5. Upper lobe of right lung, 6. Ascending aorta, 7. Superior vena cava, 8. Oblique fissure, 9. Pericardium, 10. Right atrium, 11. Lower lobe of right lung, 12. Right ventricle, 13. Left ventricle, 14. Pulmonary trunk, 15. Aortic arch, 16. Left lung, 17. Left brachiocephalic vein, 18. Left subclavian artery, 19. Left common carotid artery, 20. Right ventricle, 21. Left ventricle, 22. Interventricular septum, 23. Left atrium, 24. Left lung, 25. Esophagus, 26. Aorta, 27. Spinal cord, 28. Rib, 29. Visceral pleura, 30. Parietal pleura, 31. Body of thoracic vertebra, 32. Azygos vein, 33. Right lung, 34. Right atrium, 35. Sternum

Digestive System

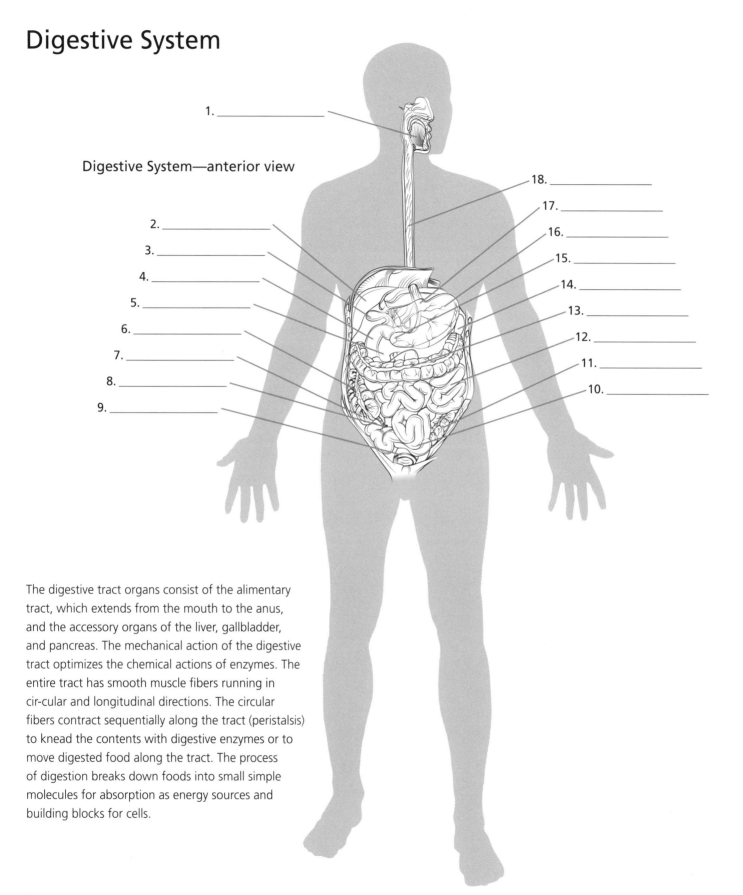

Digestive System—anterior view

1. _____

2. _____
3. _____
4. _____
5. _____
6. _____
7. _____
8. _____
9. _____

18. _____
17. _____
16. _____
15. _____
14. _____
13. _____
12. _____
11. _____
10. _____

The digestive tract organs consist of the alimentary tract, which extends from the mouth to the anus, and the accessory organs of the liver, gallbladder, and pancreas. The mechanical action of the digestive tract optimizes the chemical actions of enzymes. The entire tract has smooth muscle fibers running in cir-cular and longitudinal directions. The circular fibers contract sequentially along the tract (peristalsis) to knead the contents with digestive enzymes or to move digested food along the tract. The process of digestion breaks down foods into small simple molecules for absorption as energy sources and building blocks for cells.

Answers

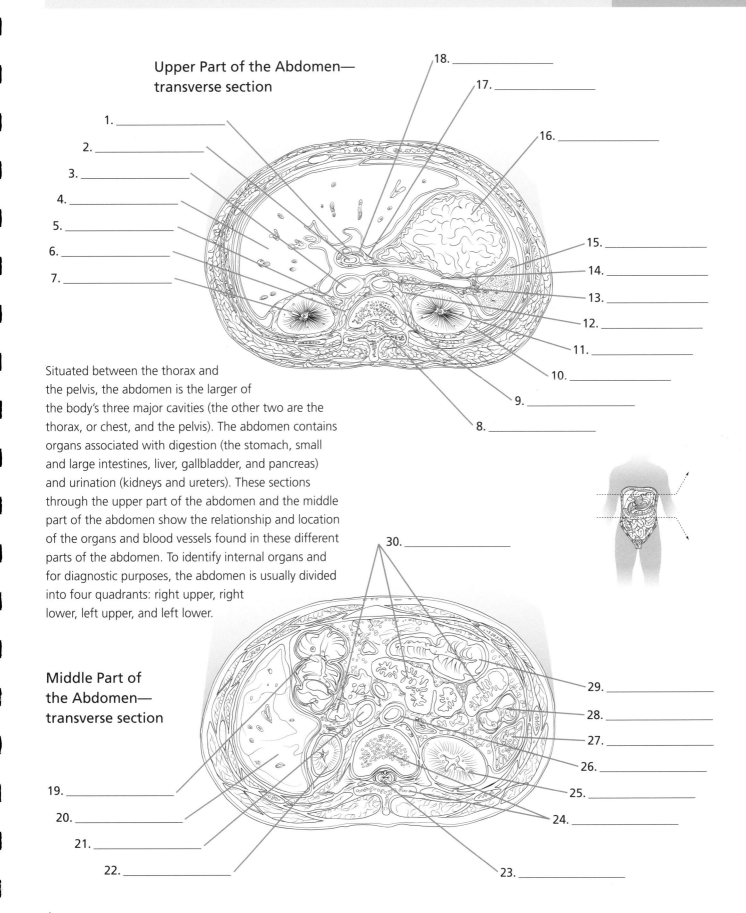

Upper Part of the Abdomen—
transverse section

18. _____

17. _____

16. _____

1. _____

2. _____

3. _____

4. _____

5. _____

6. _____

7. _____

15. _____

14. _____

13. _____

12. _____

11. _____

10. _____

9. _____

8. _____

Situated between the thorax and
the pelvis, the abdomen is the larger of
the body's three major cavities (the other two are the
thorax, or chest, and the pelvis). The abdomen contains
organs associated with digestion (the stomach, small
and large intestines, liver, gallbladder, and pancreas)
and urination (kidneys and ureters). These sections
through the upper part of the abdomen and the middle
part of the abdomen show the relationship and location
of the organs and blood vessels found in these different
parts of the abdomen. To identify internal organs and
for diagnostic purposes, the abdomen is usually divided
into four quadrants: right upper, right
lower, left upper, and left lower.

Middle Part of
the Abdomen—
transverse section

30. _____

29. _____

28. _____

27. _____

26. _____

25. _____

24. _____

19. _____

20. _____

21. _____

22. _____

23. _____

Answers

1. Common bile duct, 2. Portal vein, 3. Inferior vena cava, 4. Liver, 5. Right adrenal (suprarenal) gland, 6. Right crus of diaphragm, 7. Right kidney, 8. Spinal cord, 9. Vertebral body, 10. Perirenal fat, 11. Left kidney, 12. Left crus of diaphragm, 13. Abdominal aorta, 14. Omental bursa, 15. Spleen, 16. Stomach, 17. Lesser omentum, 18. Hepatic artery, 19. Ascending colon, 20. Liver (right lobe), 21. Inferior vena cava, 22. Right kidney, 23. Spinal cord, 24. Vertebra (body and lamina with spine), 25. Left kidney, 26. Abdominal aorta, 27. Spleen, 28. Descending colon, 29. Transverse colon, 30. Loops of small intestine

Digestive System

The oral cavity includes the mouth and its associated structures—the soft and hard palates, teeth and gums, tongue, palatine tonsils, and salivary glands. In the adult, 32 permanent teeth are arranged in two arcades of 16 teeth each. The tongue is a muscular organ attached to the floor of the mouth. On each side, the tongue is joined to the palate by the palatoglossal arches, and immediately behind these arches are the palatine tonsils. The principal salivary glands are the parotid, submandibular, and sublingual glands, which drain into the mouth through separate ducts. Lining the abdominal cavity, and also extending out into the cavity to cover the organs within, is the thin lubricating membrane of the peritoneum. Folds of the peritoneum attach the organs to the back of the abdominal cavity. Other folds of the peritoneum—the mesenteries and omenta—provide a route for the passage of nerves, blood vessels, and lymphatics.

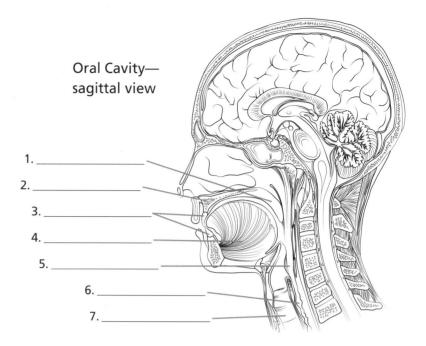

Oral Cavity—
sagittal view

1. _____
2. _____
3. _____
4. _____
5. _____
6. _____
7. _____

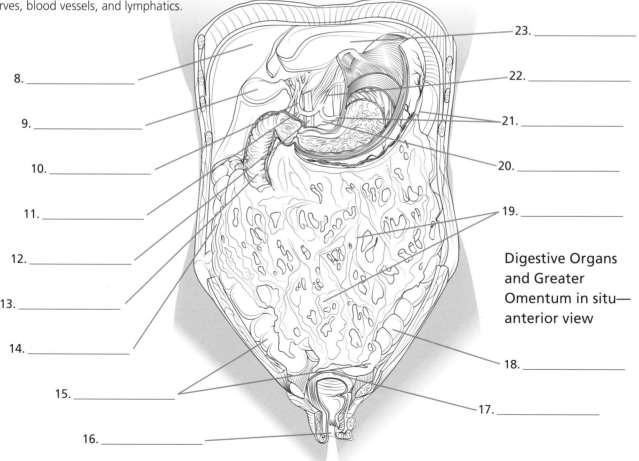

8. _____
9. _____
10. _____
11. _____
12. _____
13. _____
14. _____
15. _____
16. _____

23. _____
22. _____
21. _____
20. _____
19. _____

Digestive Organs
and Greater
Omentum in situ—
anterior view

18. _____
17. _____

Answers

Oral Cavity

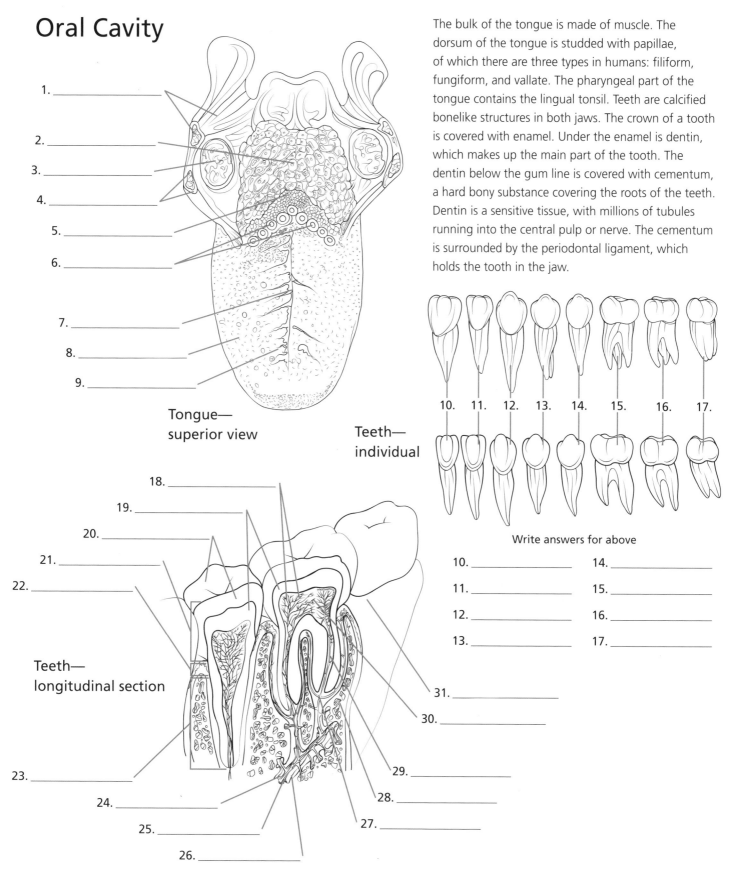

1. _____
2. _____
3. _____
4. _____
5. _____
6. _____
7. _____
8. _____
9. _____

Tongue—
superior view

The bulk of the tongue is made of muscle. The dorsum of the tongue is studded with papillae, of which there are three types in humans: filiform, fungiform, and vallate. The pharyngeal part of the tongue contains the lingual tonsil. Teeth are calcified bonelike structures in both jaws. The crown of a tooth is covered with enamel. Under the enamel is dentin, which makes up the main part of the tooth. The dentin below the gum line is covered with cementum, a hard bony substance covering the roots of the teeth. Dentin is a sensitive tissue, with millions of tubules running into the central pulp or nerve. The cementum is surrounded by the periodontal ligament, which holds the tooth in the jaw.

Teeth—
individual

10.   11.   12.   13.   14.   15.   16.   17.

Write answers for above

10. _____   14. _____
11. _____   15. _____
12. _____   16. _____
13. _____   17. _____

18. _____
19. _____
20. _____
21. _____
22. _____

Teeth—
longitudinal section

23. _____
24. _____
25. _____
26. _____
27. _____
28. _____
29. _____
30. _____
31. _____

Answers

1. Palatopharyngeal arch and muscle, 2. Lingual tonsil, 3. Palatine tonsil, 4. Palatoglossal arch and muscle, 5. Terminal sulcus, 6. Vallate papillae, 7. Median sulcus, 8. Filiform papillae, 9. Fungiform papillae, 10. Central incisor, 11. Lateral incisor, 12. Canine, 13. First premolar, 14. Second premolar, 15. Third molar, 16. Second molar, 17. First molar, 18. Capillary plexus, 19. Dentin, 20. Enamel, 21. Crown of tooth, 22. Neck of tooth, 23. Root of tooth, 24. Alveolar process, 25. Alveolar vein, 26. Alveolar artery, 27. Apical foramen, 28. Alveolar nerve, 29. Alveolar process, 30. Root canal, 31. Gingiva

Oral Cavity

The underside of the tongue is soft and is kept moist by salivary gland secretions. The salivary glands are located around the oral cavity and are divided into two groups. The major salivary glands consist of three pairs of glands: the parotid, submandibular, and sublingual glands. The minor buccal salivary glands are microscopic and are scattered around the mouth, palate, and throat. The parotid gland is located in front of each ear. The submandibular gland is located below the jaw (mandible) on each side of the neck, about 1 inch (2–3 centimeters) in front of the angle of the jaw. The sublingual glands are small and lie within ridges on the floor of the mouth, beneath the tongue. The frenulum forms a midline ridge on the lower surface of the tongue. Beneath this ridge lie paired deep arteries and veins of the tongue.

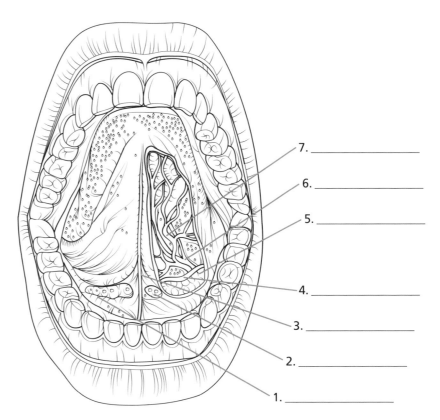

7. _____

6. _____

5. _____

4. _____

3. _____

2. _____

1. _____

Oral Cavity and Salivary Glands—
anterior view

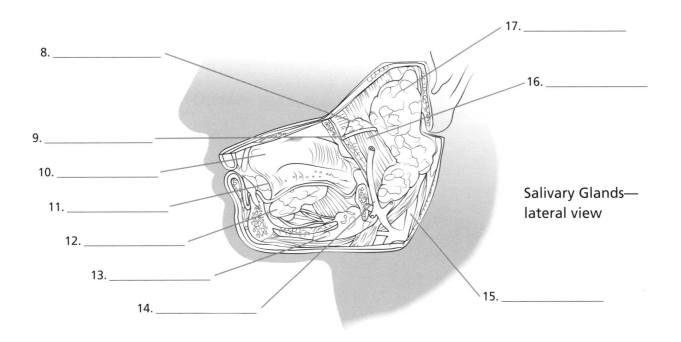

8. _____

9. _____

10. _____

11. _____

12. _____

13. _____

14. _____

17. _____

16. _____

15. _____

Salivary Glands—
lateral view

Answers

1. _____

2. _____

3. _____

8. _____

7. _____

6. _____

5. _____

4. _____

Parotid Gland
(cellular level)

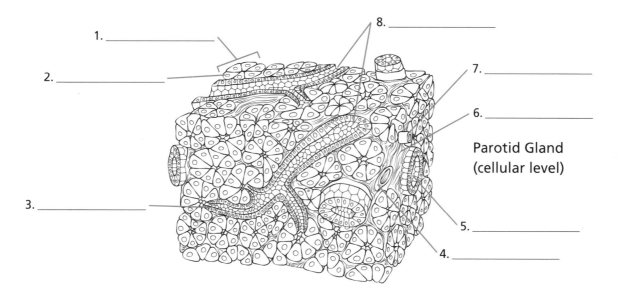

The salivary glands secrete saliva into the mouth. This fluid is needed to moisten food to ease swallowing and to begin food breakdown. The three pairs of salivary glands are the parotid, sublingual, and sub-mandibular glands. Each pair has a unique cellular organization and produces saliva with slightly different properties. The thin saliva produced by the parotid glands contains enzymes specially designed to break down starch. Each submandibular gland comprises a mixture of enzyme-producing serous cells and mucus-producing cells. Its saliva is pre-dominantly water. This gland produces thicker, watery mucus, particularly in response to milk or cream, which helps to lubricate the mouth.

Submandibular Gland
(cellular level)

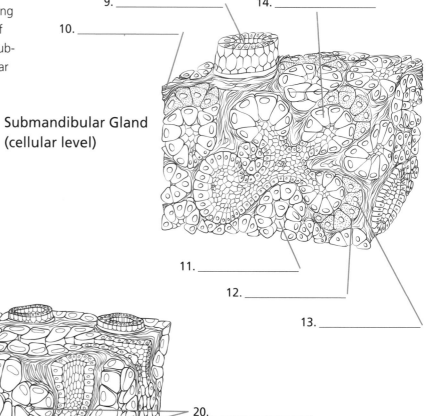

9. _____

10. _____

14. _____

11. _____

12. _____

13. _____

Sublingual Gland
(cellular level)

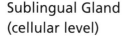

15. _____

16. _____

17. _____

20. _____

19. _____

18. _____

Answers

Liver and Gallbladder

The liver is the heaviest internal organ in the body, and it is found under the cover and protection of the lower ribs, on the right side of the abdomen. The liver has an upper (diaphragmatic) surface and a lower (visceral) surface; the two surfaces are separated at the front by a sharp inferior border. The liver is attached to the diaphragm by the falciform, triangular, and coronary ligaments. The liver is also joined to the stomach and duodenum by the gastrohepatic and hepato-duodenal ligaments, respectively. The visceral surface of the liver is in contact with the gallbladder, the right kidney, part of the duodenum, the esophagus, the stomach, and the hepatic flexure of the colon. The porta hepatis—the point where vessels and ducts enter and exit the liver—lies on the visceral surface. Microscopically, the liver contains sheets of cells (hepatocytes) arranged in hexagonal prism-shaped lobules. The space between the sheets of hepatocytes is filled with small blood vessels called liver sinusoids. Branches of both the portal vein and hepatic artery feed into these sinusoids. A system of bile ductules runs between the hepatocytes. These ductules carry bile produced by the hepatocytes. The ductules eventually join together to form intrahepatic bile ducts, which in turn join together to form the bile duct. At the corners of each hexagonal liver lobule lie a branch of the portal vein, hepatic artery, and bile duct, while the center of each lobule is occupied by a central vein.

Liver—anterior view

6. _____
5. _____
4. _____
3. _____
2. _____
1. _____

17. _____
16. _____
15. _____
7. _____
8. _____
9. _____
10. _____
11. _____
12. _____
13. _____
14. _____

Liver—posterior view

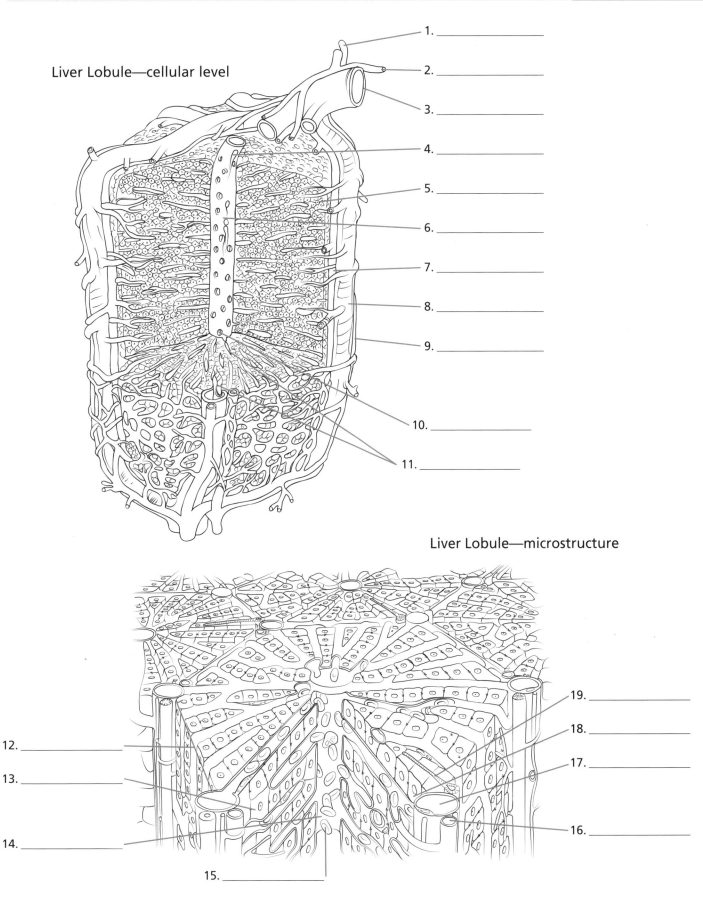

Liver Lobule—cellular level

1. _____
2. _____
3. _____
4. _____
5. _____
6. _____
7. _____
8. _____
9. _____
10. _____
11. _____

Liver Lobule—microstructure

19. _____
18. _____
17. _____
16. _____

12. _____
13. _____
14. _____
15. _____

Answers

1. Bile duct, 2. Artery, 3. Collecting vein, 4. Sublobular (interlobular) vein, 5. Hepatocyte, 6. Central vein, 7. Interlobular bile duct, 8. Branch of portal vein, 9. Branch of hepatic artery, 10. Opening of a liver sinusoid, 11. Liver sinusoids, 12. Hepatocyte plate, 13. Hepatocyte, 14. Central vein, 15. Red blood cell, 16. Hepatic artery branch, 17. Portal vein branch, 18. Sinusoid, 19. Bile canaliculus

Liver and Gallbladder

The gallbladder is a sac-shaped organ attached to the lower surface of the liver and lies on the right side of the abdomen. The gallbladder stores bile and is part of the biliary tree. The duct of the gallbladder is called the cystic duct. The hepatic ducts carry bile from the liver either to the gallbladder for storage or to the duodenum for usage. The left and right hepatic ducts join to form the common bile duct, which transports bile to the duodenum. The cystic duct transports bile from the gallbladder to the common bile duct. The portal system is the network of veins connecting the major components of the digestive tract to the liver. The primary veins in this network are the superior mesenteric, splenic, and inferior mesenteric veins, which join to form the portal vein. Also draining into the portal vein are the left gastric vein and the cystic veins.

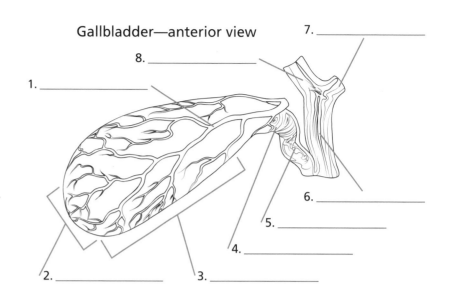

Gallbladder—anterior view

1. _____
2. _____
3. _____
4. _____
5. _____
6. _____
7. _____
8. _____

9. _____

Bile Ducts—anterior view

10. _____
11. _____

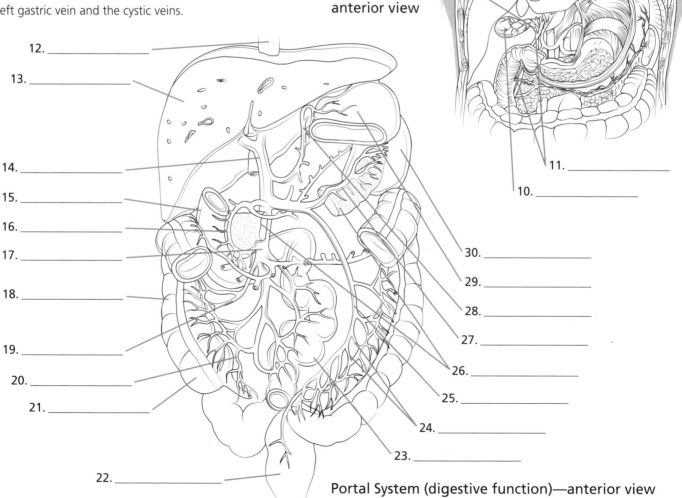

12. _____
13. _____
14. _____
15. _____
16. _____
17. _____
18. _____
19. _____
20. _____
21. _____
22. _____
23. _____
24. _____
25. _____
26. _____
27. _____
28. _____
29. _____
30. _____

Portal System (digestive function)—anterior view

Answers

Pancreas, Stomach, and Intestines

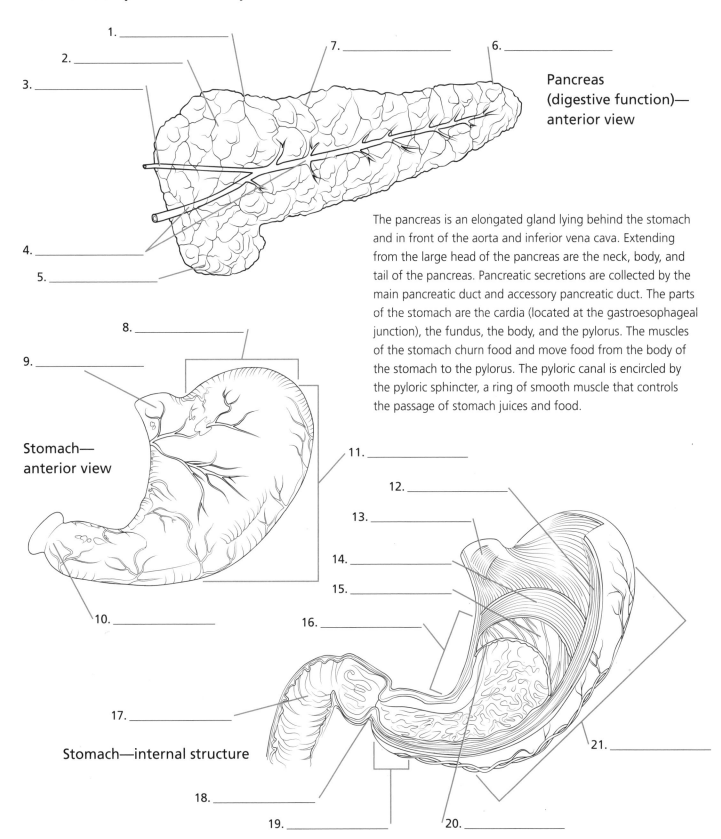

1. _____
2. _____
3. _____
7. _____
6. _____

Pancreas (digestive function)— anterior view

4. _____
5. _____

The pancreas is an elongated gland lying behind the stomach and in front of the aorta and inferior vena cava. Extending from the large head of the pancreas are the neck, body, and tail of the pancreas. Pancreatic secretions are collected by the main pancreatic duct and accessory pancreatic duct. The parts of the stomach are the cardia (located at the gastroesophageal junction), the fundus, the body, and the pylorus. The muscles of the stomach churn food and move food from the body of the stomach to the pylorus. The pyloric canal is encircled by the pyloric sphincter, a ring of smooth muscle that controls the passage of stomach juices and food.

8. _____
9. _____

Stomach— anterior view

10. _____

11. _____
12. _____
13. _____
14. _____
15. _____
16. _____

17. _____

Stomach—internal structure

18. _____
19. _____
20. _____
21. _____

Pancreas, Stomach, and Intestines

The stomach and small and large intestines are the principal organs involved in the breakdown and processing of food. The intestines occupy the lower two-thirds of the abdominal cavity and consist of two parts: the small intestine (duodenum, jejunum, and ileum) and the large intestine (colon and rectum). The small intestine leads on from the stomach, and the large intestine leads on from the small intestine. The large intestine is composed of an initial part called the cecum, with the vermiform appendix attached, and the ascending colon, transverse colon, descending colon, sigmoid colon, and rectum. The small and large intestines are covered by smooth muscle, with an inner circular layer that is thicker than the outer longitudinal layer. The lining of the intestines has many small transverse folds (plicae) and tiny projections (villi). The number of villi diminishes from the ileum to the rectum.

Jejunum—cross-sectional view

11. _____
10. _____
9. _____
8. _____
1. _____
2. _____
3. _____
4. _____
5. _____
6. _____
7. _____

12. _____
13. _____
14. _____
15. _____
16. _____
17. _____
18. _____

24. _____
23. _____
22. _____
21. _____
20. _____
19. _____

Stomach and Intestines—anterior view

Answers

1. Serosa (mesothelium), 2. Serosa (connective tissue), 3. Plicae circulares, 4. Submucosa, 5. Muscularis externa (outer longitudinal fibers), 6. Mucosa, 7. Muscularis mucosae, 8. Nerves of the mesenteric plexus, 9. Muscularis externa (inner circular layer), 10. Villus, 11. Mesentery, 12. Duodenum, 13. Transverse colon (reflected down), 14. Ascending colon, 15. Cecum, 16. Appendix, 17. Rectum, 18. Anus, 19. Ileum, 20. Sigmoid colon, 21. Jejunum, 22. Pylorus (of stomach), 23. Ascending colon, 24. Stomach

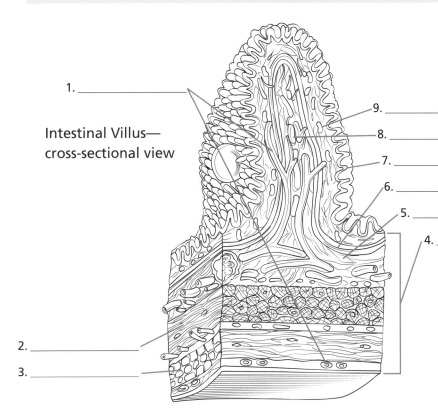

Intestinal Villus— cross-sectional view

1. _____

9. _____

8. _____

7. _____

6. _____

5. _____

4. _____

2. _____

3. _____

The plicae and villi on the lining of the duodenum and jejunum greatly increase the surface area available for the absorption of nutrients. The rectum is the second-to-last part of the digestive tract. Its upper part has a series of folds in its walls called rectal valves. At the lower end of the rectum are longitudinally running folds called anal (or rectal) columns. The anus is a short tube leading from the rectum through the anal sphincter to the anal orifice. The anus is closed by an involuntary and a voluntary anal sphincter—the internal and external anal sphincter, respectively.

Anus—coronal view

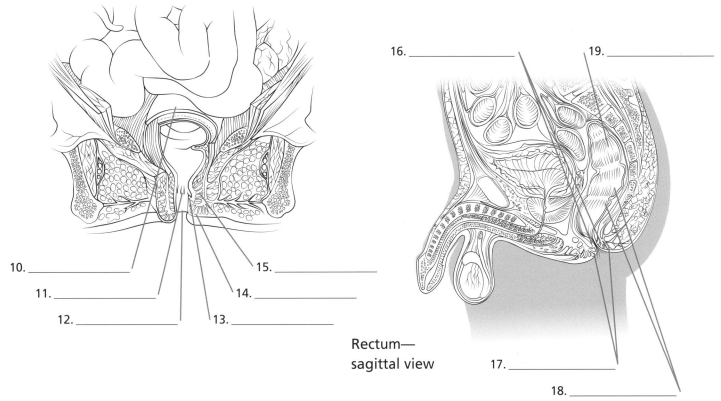

10. _____

11. _____

12. _____

15. _____

14. _____

13. _____

16. _____

19. _____

Rectum— sagittal view

17. _____

18. _____

Answers

Urinary System

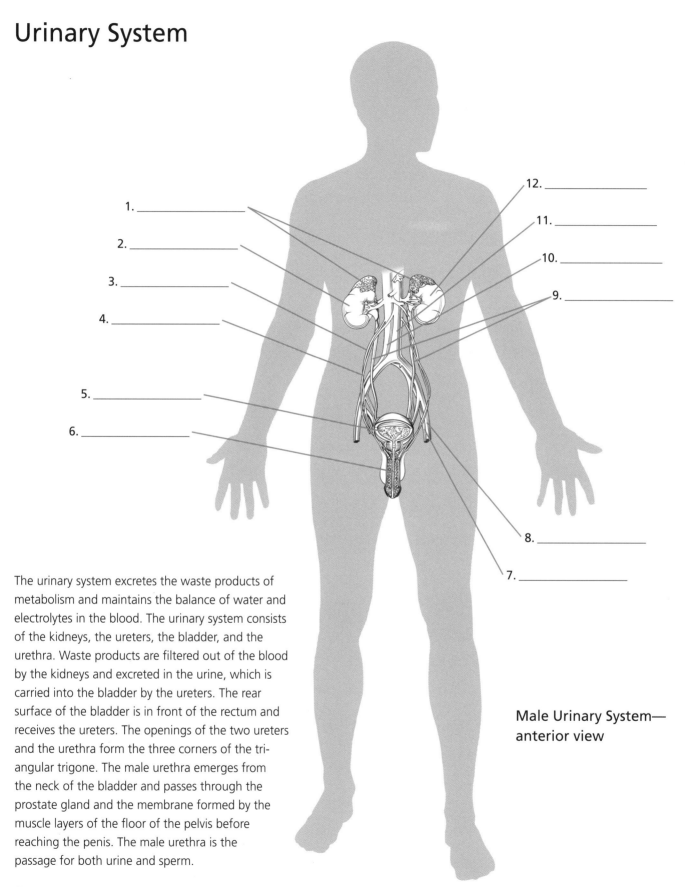

1. _____

2. _____

3. _____

4. _____

5. _____

6. _____

12. _____

11. _____

10. _____

9. _____

8. _____

7. _____

The urinary system excretes the waste products of metabolism and maintains the balance of water and electrolytes in the blood. The urinary system consists of the kidneys, the ureters, the bladder, and the urethra. Waste products are filtered out of the blood by the kidneys and excreted in the urine, which is carried into the bladder by the ureters. The rear surface of the bladder is in front of the rectum and receives the ureters. The openings of the two ureters and the urethra form the three corners of the tri-angular trigone. The male urethra emerges from the neck of the bladder and passes through the prostate gland and the membrane formed by the muscle layers of the floor of the pelvis before reaching the penis. The male urethra is the passage for both urine and sperm.

Male Urinary System—
anterior view

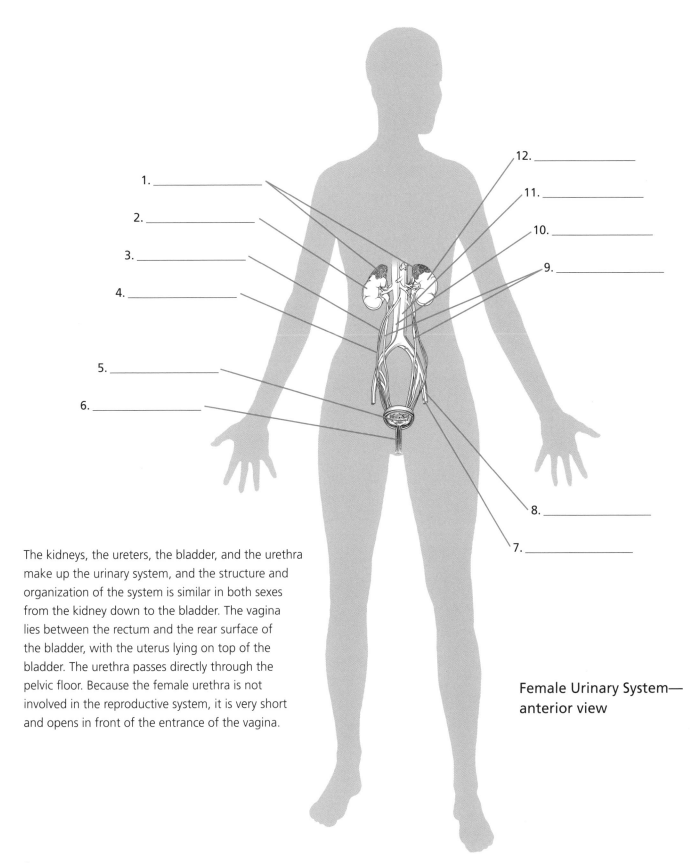

1. _____

2. _____

3. _____

4. _____

5. _____

6. _____

12. _____

11. _____

10. _____

9. _____

8. _____

7. _____

The kidneys, the ureters, the bladder, and the urethra make up the urinary system, and the structure and organization of the system is similar in both sexes from the kidney down to the bladder. The vagina lies between the rectum and the rear surface of the bladder, with the uterus lying on top of the bladder. The urethra passes directly through the pelvic floor. Because the female urethra is not involved in the reproductive system, it is very short and opens in front of the entrance of the vagina.

Female Urinary System— anterior view

Answers

1. Adrenal glands, 2. Right kidney, 3. Ovarian vein, 4. Ovarian artery, 5. Bladder, 6. Urethra, 7. External iliac vein, 8. External iliac artery, 9. Ureters, 10. Inferior vena cava, 11. Abdominal aorta, 12. Left kidney

Kidneys and Urinary Tract

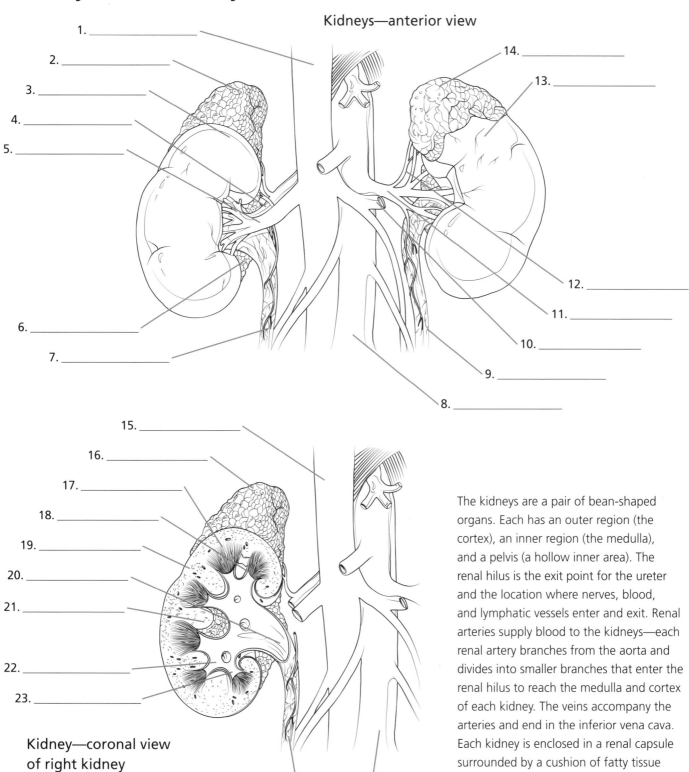

Kidneys—anterior view

1. _____

2. _____

3. _____

4. _____

5. _____

6. _____

7. _____

14. _____

13. _____

12. _____

11. _____

10. _____

9. _____

8. _____

15. _____

16. _____

17. _____

18. _____

19. _____

20. _____

21. _____

22. _____

23. _____

Kidney—coronal view
of right kidney

24. _____

25. _____

The kidneys are a pair of bean-shaped organs. Each has an outer region (the cortex), an inner region (the medulla), and a pelvis (a hollow inner area). The renal hilus is the exit point for the ureter and the location where nerves, blood, and lymphatic vessels enter and exit. Renal arteries supply blood to the kidneys—each renal artery branches from the aorta and divides into smaller branches that enter the renal hilus to reach the medulla and cortex of each kidney. The veins accompany the arteries and end in the inferior vena cava. Each kidney is enclosed in a renal capsule surrounded by a cushion of fatty tissue and a layer of connective tissue.

Answers

Urinary Tract—anterior view

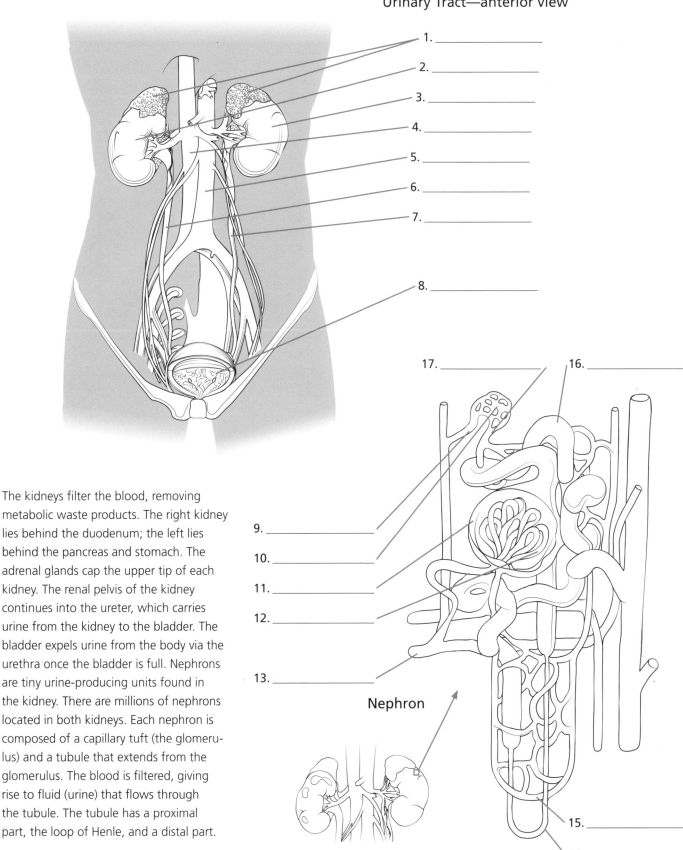

1. _____
2. _____
3. _____
4. _____
5. _____
6. _____
7. _____
8. _____

17. _____   16. _____

9. _____
10. _____
11. _____
12. _____
13. _____

Nephron

15. _____
14. _____

The kidneys filter the blood, removing metabolic waste products. The right kidney lies behind the duodenum; the left lies behind the pancreas and stomach. The adrenal glands cap the upper tip of each kidney. The renal pelvis of the kidney continues into the ureter, which carries urine from the kidney to the bladder. The bladder expels urine from the body via the urethra once the bladder is full. Nephrons are tiny urine-producing units found in the kidney. There are millions of nephrons located in both kidneys. Each nephron is composed of a capillary tuft (the glomerulus) and a tubule that extends from the glomerulus. The blood is filtered, giving rise to fluid (urine) that flows through the tubule. The tubule has a proximal part, the loop of Henle, and a distal part.

Answers

1. Adrenal glands, 2. Right kidney, 3. Left kidney, 4. Inferior vena cava, 5. Abdominal aorta, 6. Right ureter, 7. Left ureter, 8. Urinary bladder, 9. Afferent arteriole, 10. Glomerular capillaries, 11. Bowman's capsule around a glomerulus, 12. Glomerulus, 13. Artery, 14. Loop of Henle, 15. Capillary network, 16. Proximal convoluted tubule, 17. Efferent arteriole

Urinary Tract and Bladder

Urine flows from the kidneys through the ureters to the bladder. The ureters are two muscular tubes, the walls of which contain smooth muscles that contract and propel urine in waves into the bladder, a muscular sac that resembles an inverted pyramid. The rear surface of the bladder receives the ureters. The urethra emerges from the neck of the bladder, which is located at the bottom of the bladder. The openings of the two ureters and the urethra form the three corners of the triangular trigone. Two muscles—internal and external urethral sphincter muscles—work in unison to control elimination of urine. The internal urethral sphincter muscle is controlled by the autonomic nervous system, alerting the brain when the bladder is full. The external urethral sphincter is under voluntary control, and only when this muscle is relaxed does the internal sphincter also relax, allowing the elimination of urine. The urethra carries urine from the bladder to the outside of the body. The male urethra is usually 8 inches (20 centimeters) long, passes from the bladder through the prostate gland, where it is joined by the sperm ducts, and opens at the tip of the penis. Essentially similar in structure to the male urinary system, the female urinary system features a shorter urethra than that of the male. Urine passes along the ureters to the bladder and then is expelled from the body via the urethra, which opens out in front of the entrance to the vagina.

Urinary Tract (male)— sagittal view

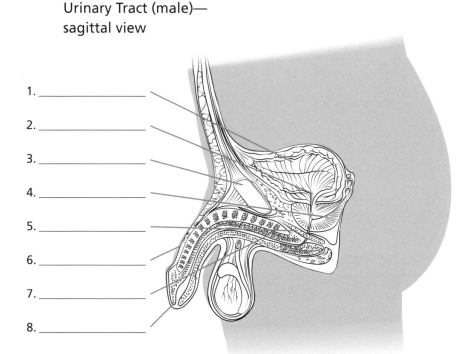

1. _____
2. _____
3. _____
4. _____
5. _____
6. _____
7. _____
8. _____

Urinary Tract (female)— sagittal view

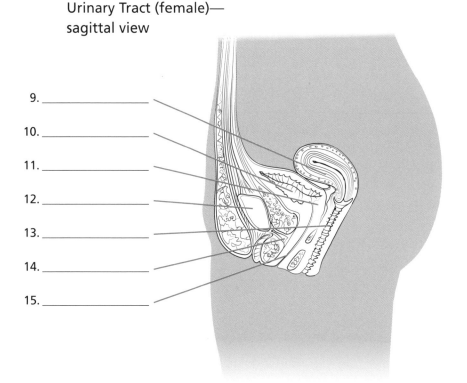

9. _____
10. _____
11. _____
12. _____
13. _____
14. _____
15. _____

Answers

1. Bladder, 2. Region of internal urethral sphincter, 3. Pubic symphysis, 4. Prostate gland, 5. External urethral sphincter, 6. Penis, 7. Urethra, 8. Urethral meatus (opening), 9. Uterus, 10. Bladder, 11. Internal urethral sphincter, 12. Pubic symphysis, 13. Vagina, 14. External urethral sphincter, 15. Urethra

Bladder (male)—
anterior view

Bladder (male)—
posterior view

Bladder (female)—
anterior view

1. _____
2. _____
3. _____
4. _____
5. _____
6. _____
7. _____
8. _____
9. _____
10. _____
11. _____

12. _____
13. _____
14. _____
15. _____
16. _____
17. _____
18. _____
19. _____
20. _____
21. _____
22. _____
23. _____
24. _____

26. _____
25. _____
36. _____
35. _____
34. _____
33. _____
32. _____
31. _____
30. _____
29. _____
28. _____
27. _____

37. _____
38. _____
39. _____
40. _____
41. _____
42. _____
43. _____

44. _____
45. _____
46. _____
47. _____
48. _____
49. _____
50. _____
51. _____
52. _____

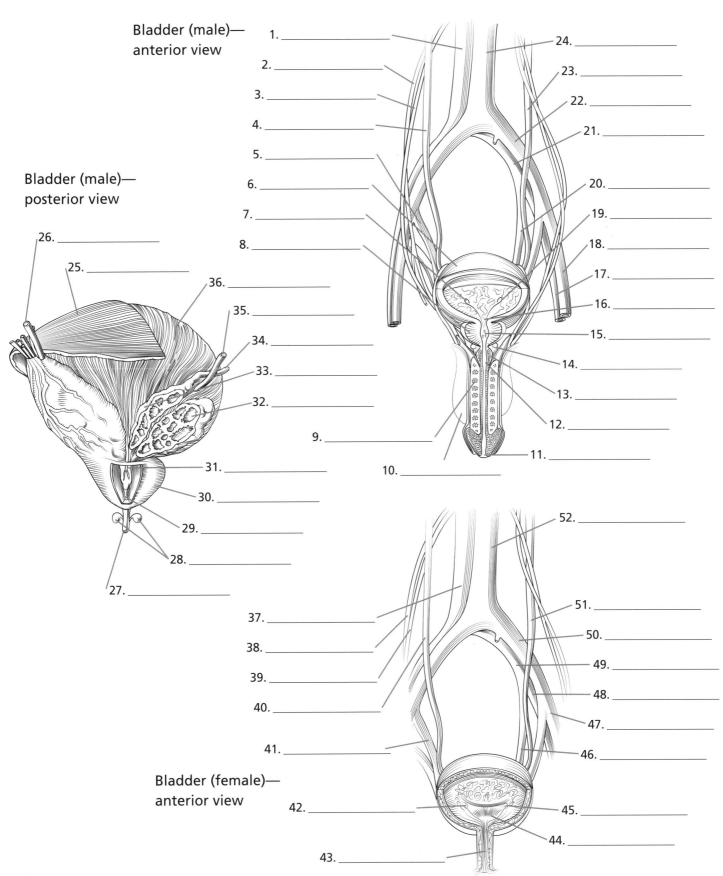

Answers

1. Inferior vena cava, 2. Right testicular vein, 3. Right testicular artery, 4. Right ureter, 5. Right internal iliac artery, 6. Bladder, 7. Opening (meatus) of right ureter, 8. Prostate gland, 9. Penis, 10. Scrotum, 11. Urethral meatus (opening), 12. Spongy urethra, 13. Bulb of penis, 14. Bulbourethral (Cowper's) gland, 15. Prostatic urethra, 16. Neck of bladder, 17. Left external iliac artery, 18. Left external iliac artery, 19. Opening (meatus) of left ureter, 20. Left internal iliac vein, 21. Left common iliac vein, 22. Left common iliac artery, 23. Left ureter, 24. Abdominal aorta, 25. Peritoneum, 26. Left ureter, 27. Membranous urethra, 28. Bulbourethral (Cowper's) glands, 29. Prostatic urethra, 30. Prostate gland, 31. Ejaculatory duct, 32. Seminal vesicle, 33. Ampulla of ductus deferens, 34. Ductus deferens, 35. Right ureter, 36. Detrusor muscle of bladder, 37. Inferior vena cava, 38. Right ovarian vein, 39. Right ovarian artery, 40. Right ureter, 41. Right internal iliac artery, 42. Opening (meatus) of right ureter, 43. Urethra, 44. Trigone, 45. Opening (meatus) of left ureter, 46. Left internal iliac vein, 47. Left external iliac vein, 48. Left external iliac artery, 49. Left common iliac vein, 50. Left common iliac artery, 51. Left ureter, 52. Abdominal aorta

Endocrine System

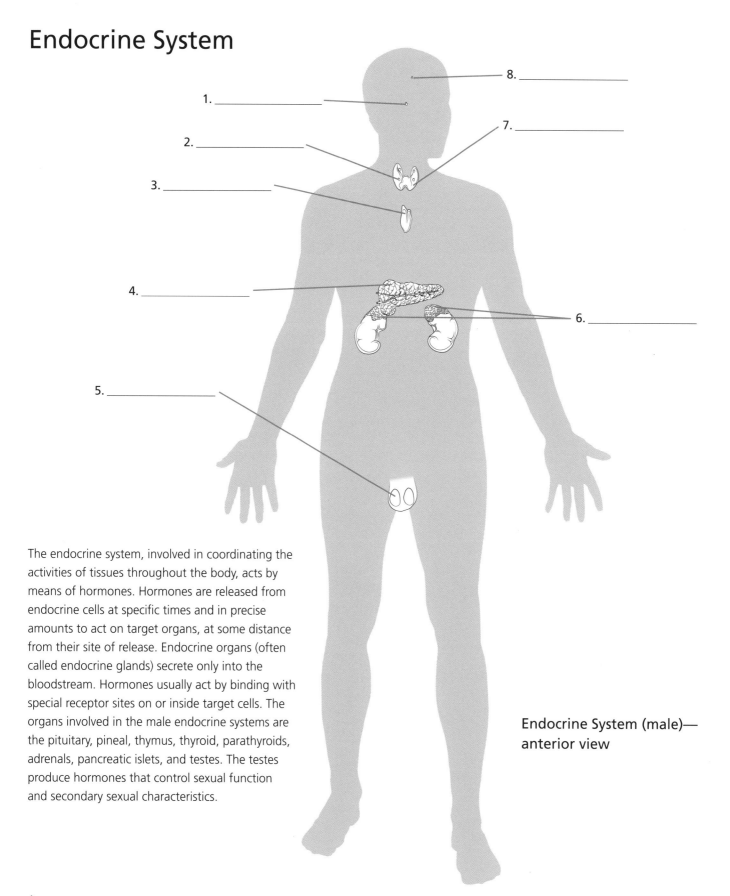

1. _____
2. _____
3. _____
4. _____
5. _____
6. _____
7. _____
8. _____

The endocrine system, involved in coordinating the activities of tissues throughout the body, acts by means of hormones. Hormones are released from endocrine cells at specific times and in precise amounts to act on target organs, at some distance from their site of release. Endocrine organs (often called endocrine glands) secrete only into the bloodstream. Hormones usually act by binding with special receptor sites on or inside target cells. The organs involved in the male endocrine systems are the pituitary, pineal, thymus, thyroid, parathyroids, adrenals, pancreatic islets, and testes. The testes produce hormones that control sexual function and secondary sexual characteristics.

Endocrine System (male)—
anterior view

Answers

1. Pituitary gland, 2. Parathyroid, 3. Parathyroid, 4. Thymus, 5. Pancreas, 5. Testes, 6. Adrenal glands, 7. Thyroid gland 8. Pineal gland

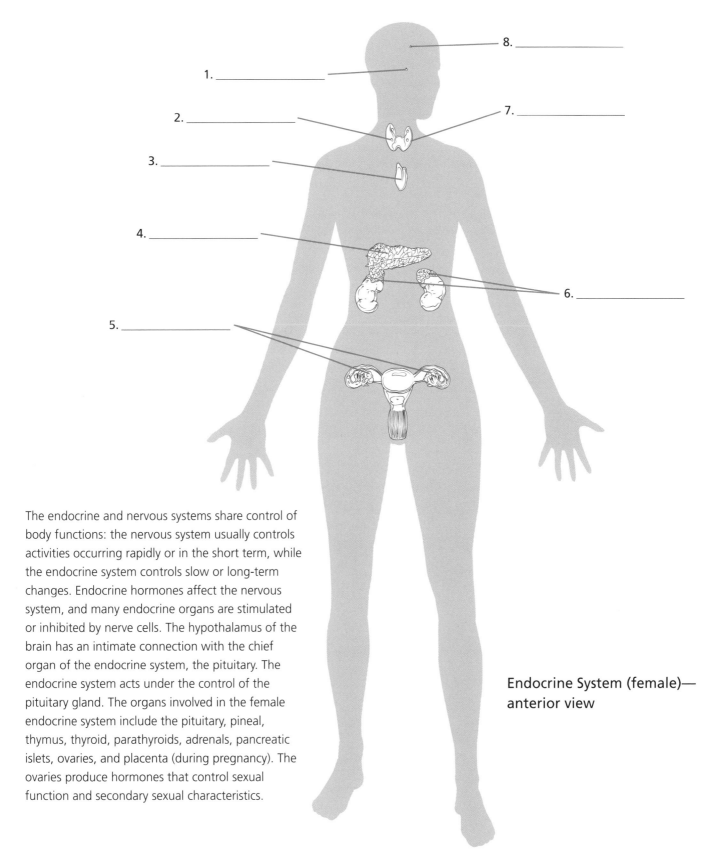

The endocrine and nervous systems share control of body functions: the nervous system usually controls activities occurring rapidly or in the short term, while the endocrine system controls slow or long-term changes. Endocrine hormones affect the nervous system, and many endocrine organs are stimulated or inhibited by nerve cells. The hypothalamus of the brain has an intimate connection with the chief organ of the endocrine system, the pituitary. The endocrine system acts under the control of the pituitary gland. The organs involved in the female endocrine system include the pituitary, pineal, thymus, thyroid, parathyroids, adrenals, pancreatic islets, ovaries, and placenta (during pregnancy). The ovaries produce hormones that control sexual function and secondary sexual characteristics.

Endocrine System (female)— anterior view

Answers

1. Pituitary gland, 2. Parathyroid, 3. Thymus, 4. Pancreas, 5. Ovaries, 6. Adrenal glands, 7. Thyroid gland 8. Pineal gland

Endocrine Glands

The pituitary gland (hypophysis) is a very small organ lying immediately below the hypothalamus of the brain. The pituitary gland is divided into anterior and posterior lobes (the adeno-hypophysis and neurohypophysis, respectively). The anterior lobe contains many different types of cells, which produce growth hormone, prolactin, follicle-stimulating hormone, luteinizing hormone, thyroid-stimulating hormone, adreno-corticotrophic hormone, and melanocyte-stimulating hormone. The posterior lobe contains oxytocin and antidiuretic hormones, produced in the hypothalamus and transported to the pituitary within nerve fibers. The pineal gland (body) is also small and is located inside the skull cavity, surrounded by the brain. The pineal gland produces melatonin, whose concentration varies in tune with the 24-hour cycle of the day (circadian rhythm)—melatonin production is highest during a person's normal sleeping hours and drops off as the body begins to wake. The pineal gland probably has an effect on the ovaries and testes and may influence mood.

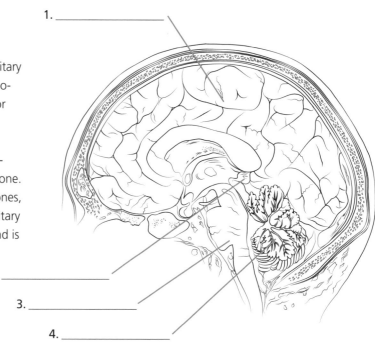

1. _____

2. _____

3. _____

4. _____

Pineal Gland

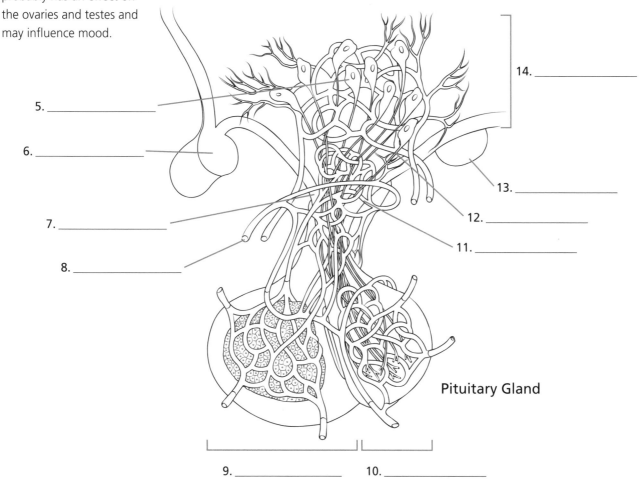

5. _____

6. _____

7. _____

8. _____

14. _____

13. _____

12. _____

11. _____

9. _____   10. _____

Pituitary Gland

Answers

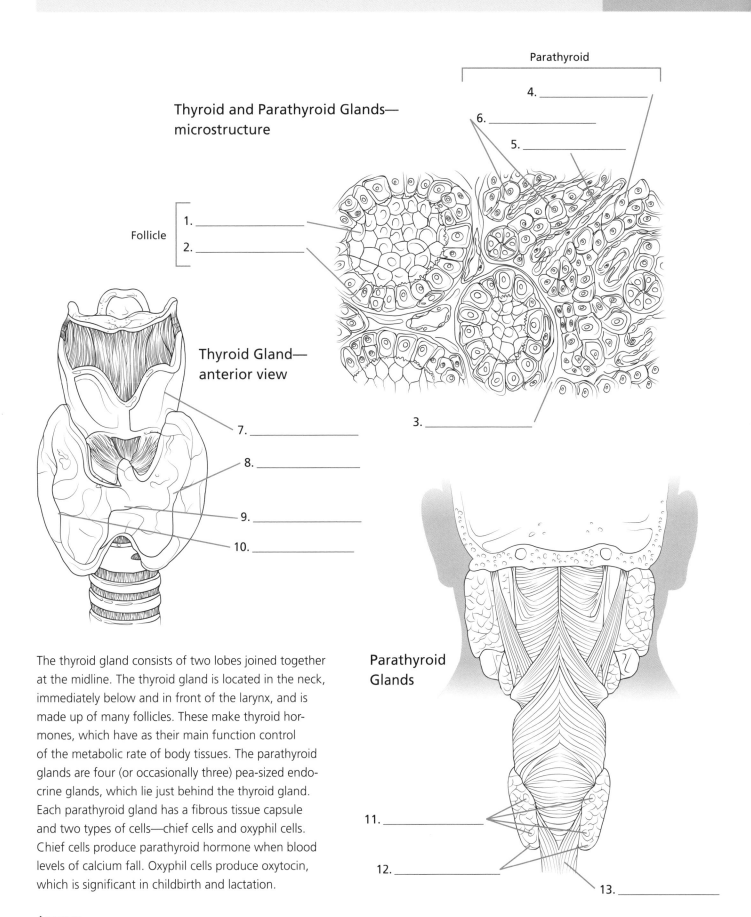

Thyroid and Parathyroid Glands— microstructure

Parathyroid

4. _____

6. _____

5. _____

Follicle

1. _____

2. _____

3. _____

Thyroid Gland— anterior view

7. _____

8. _____

9. _____

10. _____

The thyroid gland consists of two lobes joined together at the midline. The thyroid gland is located in the neck, immediately below and in front of the larynx, and is made up of many follicles. These make thyroid hormones, which have as their main function control of the metabolic rate of body tissues. The parathyroid glands are four (or occasionally three) pea-sized endocrine glands, which lie just behind the thyroid gland. Each parathyroid gland has a fibrous tissue capsule and two types of cells—chief cells and oxyphil cells. Chief cells produce parathyroid hormone when blood levels of calcium fall. Oxyphil cells produce oxytocin, which is significant in childbirth and lactation.

Parathyroid Glands

11. _____

12. _____

13. _____

Answers

Endocrine Glands

The pancreas is an elongated gland lying behind the stomach. The pancreas plays a role in both the digestive and the endocrine systems. Its endocrine function is hormonal and is involved in regulation of glucose mobilization and storage (the responsible hormones are glucagon and insulin, respectively). These two hormones are produced in special cell types (alpha and beta cells) within many tiny spherical clumps of pancreatic tissue, known as pancreatic islets or the islets of Langerhans. The islets are surrounded by exocrine tissue.

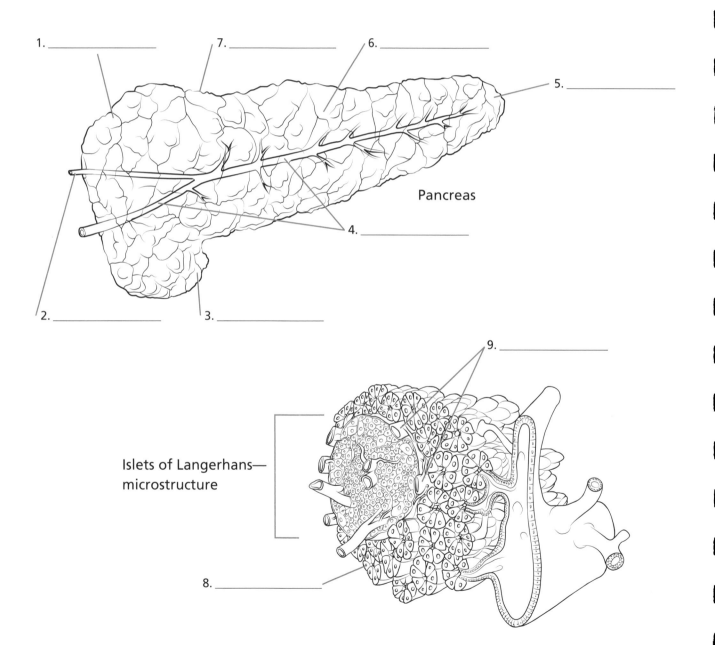

1. _____

7. _____

6. _____

5. _____

Pancreas

2. _____

3. _____

4. _____

9. _____

Islets of Langerhans—
microstructure

8. _____

Answers

The two triangular adrenal (or suprarenal) glands lie one on top of each kidney at the back of the abdomen. Each gland has an outer part (the cortex) and a core (the medulla). The adrenal cortex produces three main types of hormones: glucocorticoids, mineralocorticoids, and sex steroids. Glucocorticoids promote the breakdown of protein and the release of fat and sugars into the bloodstream. Mineralocorticoids stimulate the release of sodium in the kidneys. Sex steroids contribute to the development of sexual characteristics. The adrenal medulla contains many modified nerve cells, which produce the hormones epinephrine and norepinephrine (adrenaline and noradrenaline), respectively. These hormones are released in bursts during emergency situations or accompanying intense emotion. They act to increase the strength and rate of heart contraction, raise the blood sugar level, elevate blood pressure, and increase breathing.

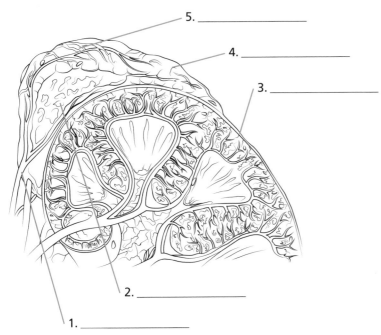

5. _____
4. _____
3. _____
2. _____
1. _____

Adrenal Glands—
coronal section through
left adrenal gland

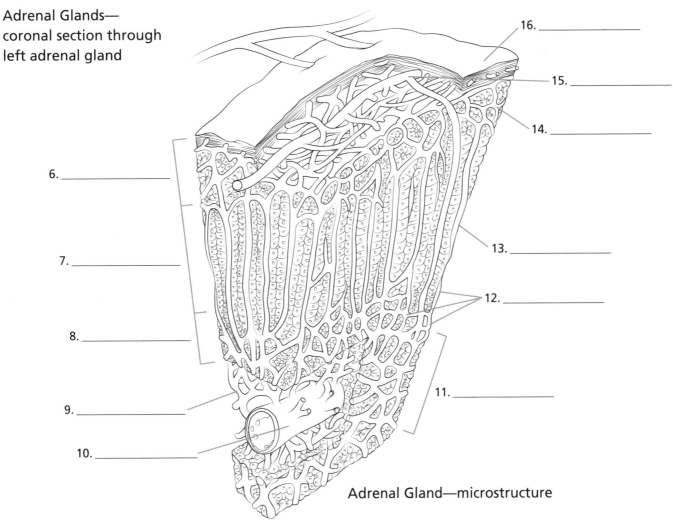

6. _____
7. _____
8. _____
9. _____
10. _____
11. _____
12. _____
13. _____
14. _____
15. _____
16. _____

Adrenal Gland—microstructure

Answers

1. Left suprarenal artery, 2. Left adrenal medulla, 3. Left kidney, 4. Left adrenal cortex, 5. Left adrenal gland, 6. Zona glomerulosa, 7. Zona fasciculata, 8. Zona reticularis, 9. Medullary plexus of veins, 10. Medullary vein, 11. Medulla, 12. Deep plexus of veins, 13. Sinusoidal vessels, 14. Subcapsular plexus of veins, 15. Capsular artery, 16. Capsule

Male and Female Endocrine Glands

The testes are two ovoid organs contained in the scrotum, a sac that lies directly behind and beneath the penis. The testes produce male sex hormones—primarily testosterone. Testosterone is the hormone responsible for the development of secondary sexual characteristics in the male— it stimulates growth of facial and pubic hair, enlargement of the larynx and deepening of the voice, and an increase in muscle tone. Tes-tosterone production is controlled by follicle-stimulating hormone (FSH) and luteinizing hormone (LH), both of which are secreted by the anterior lobe of the pituitary gland. The two ovaries are the female gonads in which the ova are formed. They resemble almonds in shape and size, and are situated on either side of the uterus. The ovaries produce the female hormones estrogen and progesterone. Estrogen causes growth of the breasts and reproductive organs, among other functions. Progesterone maintains the lining of the uterus in a state suitable to receive a fertilized ovum.

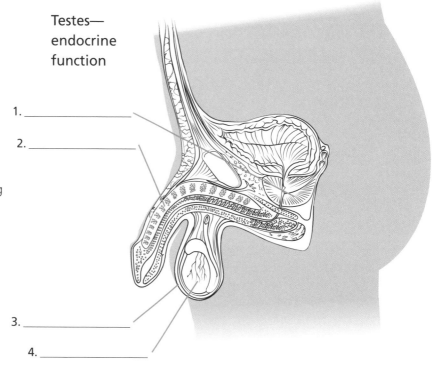

Testes— endocrine function

1. _____

2. _____

3. _____

4. _____

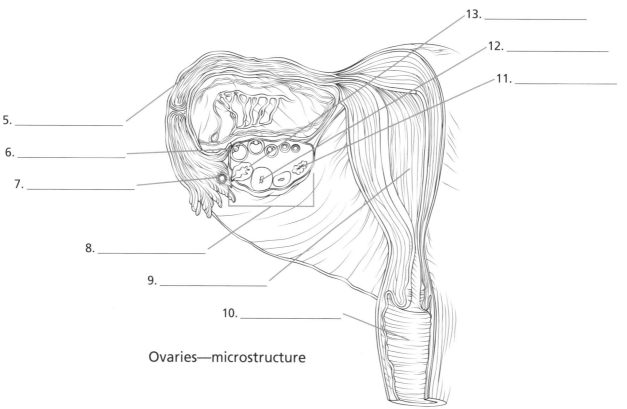

13. _____

12. _____

11. _____

5. _____

6. _____

7. _____

8. _____

9. _____

10. _____

Ovaries—microstructure

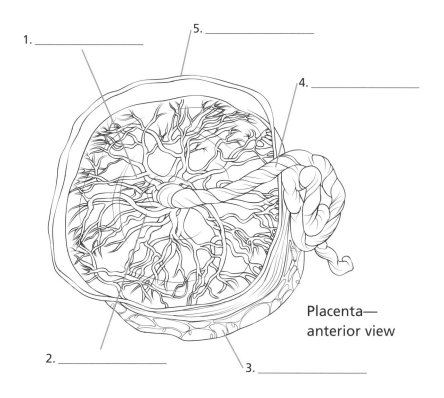

1. _____

5. _____

4. _____

2. _____

3. _____

Placenta—anterior view

An organ of pregnancy, the placenta connects the baby to the mother via the umbilical cord. The placenta also acts as an endocrine organ, producing a hormone to sustain the pregnancy (human chorionic gonadotrophin). HCG is the earliest placental hormone to be produced and is first secreted on day 6 of gestation. HCG maintains the corpus luteum (the ovarian follicle from which the ovum burst) and ensures that it continues to manufacture progesterone and estrogen until the placenta is able to produce adequate amounts of both, usually by the third month of gestation, when HCG levels decline. The placenta also produces estrogen, progesterone and relaxin (which relaxes the pelvic ligaments), along with human placental lactogen, which promotes milk production and fetal growth.

Placenta—cross-sectional view

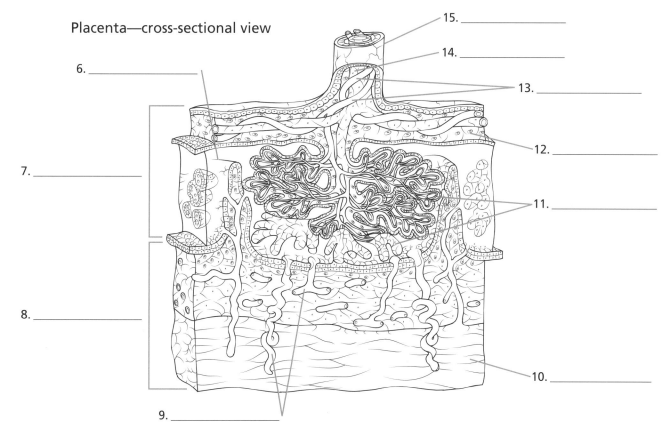

15. _____

14. _____

13. _____

6. _____

12. _____

11. _____

7. _____

8. _____

10. _____

9. _____

Answers

Male Reproductive System and Organs

The male reproductive system includes two testes, which produce spermatozoa and male hormones, a system of ducts that convey sperm, glands that contribute secretions to semen, and the external genitalia—the scrotum and penis. Sperm are formed in the testes. During ejaculation, the sperm combine with secretions from the prostate gland and seminal vesicles to form the seminal fluid. The testes are two oval organs contained in the scrotum. Each testis is capped by the epididymis, which becomes the ductus deferens. The ductus deferens and its surrounding vessels and nerves form the contents of the spermatic cord. The ductus deferens then joins the duct of the seminal vesicle to form the ejaculatory duct. The ejaculatory duct opens into the prostatic urethra. The urethra enters and courses through the penis.

1. _____

2. _____

3. _____

4. _____

Male Reproductive System—anterior view

Male Reproductive System— sagittal view

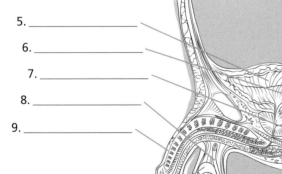

5. _____

6. _____

7. _____

8. _____

9. _____

10. _____

11. _____

Answers

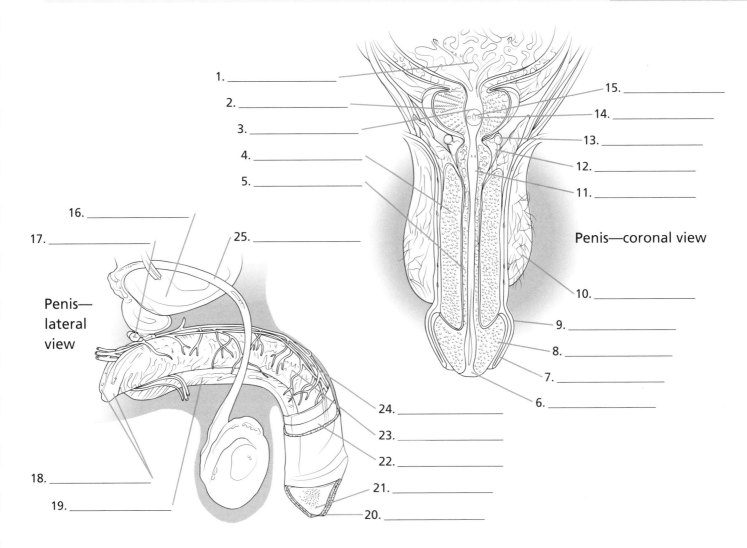

1. _____

2. _____

3. _____

4. _____

5. _____

15. _____

14. _____

13. _____

12. _____

11. _____

Penis—coronal view

10. _____

9. _____

8. _____

7. _____

6. _____

16. _____

17. _____

Penis— lateral view

25. _____

24. _____

23. _____

22. _____

21. _____

20. _____

18. _____

19. _____

The penis is the external male reproductive and urinary organ through which semen and urine leave the body, and it is attached at its base to the pelvic bone by connective tissue. The penis is composed primarily of two cylinders of spongelike vascular tissue (the corpora cavernosa) and a third cylinder (the corpus spongiosum) that contains the urethra, which carries both ejaculate and urine. The urethra ends at the glans, an external swelling at the tip of the penis. The glans is covered by the prepuce (foreskin). The prostate gland is shaped like an inverted pyramid and lies under the bladder, with the apex pointing downward. All the glands of the prostate open into the prostatic urethra and secrete the enzyme acid phosphatase, fibrinolysin, and other proteins.

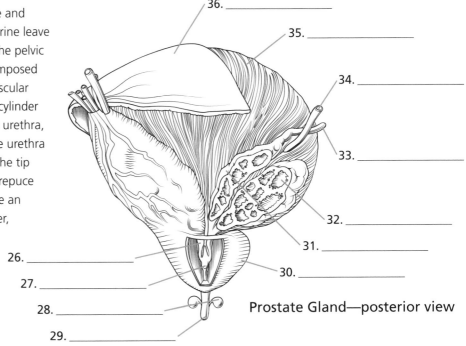

36. _____

35. _____

34. _____

33. _____

32. _____

31. _____

30. _____

26. _____

27. _____

28. _____

29. _____

Prostate Gland—posterior view

Answers

1. Bladder, 2. Prostate gland, 3. Colliculus seminalis, 4. Corpus cavernosum, 5. Corpus spongiosum, 6. Urethral meatus, 7. Prepuce, 8. Glans of penis, 9. Corona glandis, 10. Scrotum, 11. Urethra, 12. Bulb of penis, 13. Bulbourethral (Cowper's) gland, 14. Opening of ejaculatory duct, 15. Prostatic utricle, 16. Prostate gland, 17. Bulbourethral (Cowper's) gland, 18. Crura of penis, 19. Corpus spongiosum, 20. Prepuce, 21. Glans of penis, 22. Fascia penis, 23. Corpus cavernosum, 24. Superficial dorsal vein, 25. Ductus deferens, 26. Ductus deferens, 27. Ejaculatory duct, 28. Bulbourethral gland, 29. Membranous urethra, 30. Prostate gland, 31. Ampulla of ductus deferens, 32. Seminal vesicle, 33. Ductus deferens, 34. Ureter, 35. Detrusor muscle (of bladder), 36. Peritoneum.

Male Reproductive System and Organs

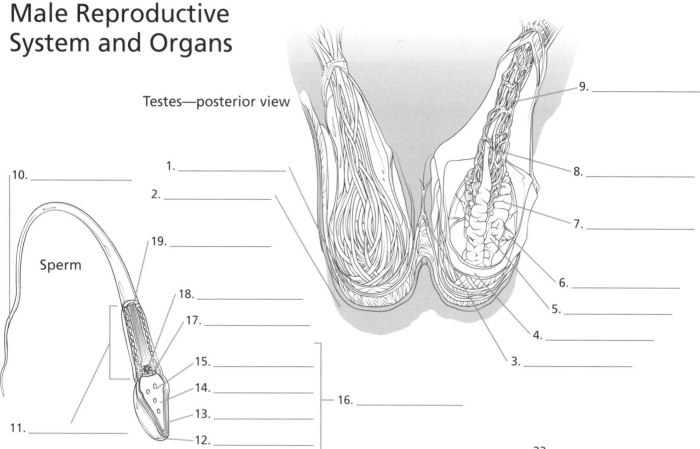

Testes—posterior view

9. _____

8. _____

7. _____

6. _____

5. _____

4. _____

3. _____

10. _____

Sperm

1. _____

2. _____

19. _____

18. _____

17. _____

15. _____

14. _____

13. _____

12. _____

16. _____

11. _____

The testes, or testicles, are two ovoid organs contained in the scrotum, a sac that lies directly posterior to the penis. A central septum of the scrotum separates the testes. Each testis has a tough inelastic fibrous capsule called the tunica albuginea that sends partitions inward to divide it into about 300 lobules. Each lobule contains coiled seminiferous tubules that converge into a network that sends about 20 small ducts through the tunica albuginea into the epididymis. The ducts become larger and convoluted and gradually fuse into the ductus deferens. Sperm are produced in the seminiferous tubules. The head of each sperm has a nucleus that contains chromosomes and an acrosomal membrane that holds enzymes needed for fertilization. The tail of the sperm helps it move in a corkscrew action on its journey from the testes to the female reproductive organs.

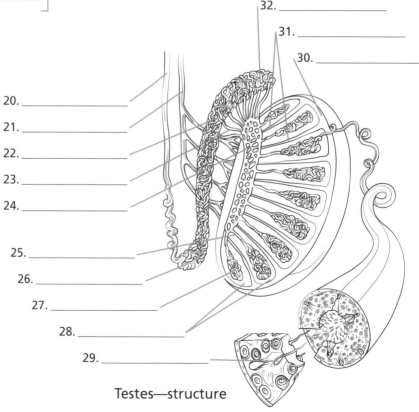

32. _____

31. _____

30. _____

20. _____

21. _____

22. _____

23. _____

24. _____

25. _____

26. _____

27. _____

28. _____

29. _____

Testes—structure

Answers

1. Superficial fascia of scrotum, 2. Scrotal skin, 3. Septum of scrotum, 4. Parietal layer of tunica vaginalis, 5. Body of epididymis, 6. Testis (covered by visceral layer of tunica vaginalis), 7. Head of epididymis, 8. Ductus deferens, 9. Pampiniform plexus of veins, 10. Tail, 11. Mitochondrial sheath (middle piece), 12. Cell membrane, 13. Acrosome, 14. Nucleus, 15. Nuclear vacuole, 16. Head, 17. Neck, 18. Centriole, 19. Mitochondrion, 20. Ductus deferens, 21. Testicular artery, 22. Body of epididymis, 23. Efferent ductules, 24. Rete testis, 25. Mediastinum testis, 26. Tail of epididymis, 27. Seminiferous tubules, 28. Lobules, 29. Spermatozoa, 30. Tunica albuginea, 31. Septae, 32. Head of epididymis

Female Reproductive System and Organs

The female reproductive system includes two ovaries that produce ova and female hormones, two uterine tubes that convey the ova to the uterus, the uterus itself, and the vagina, which is connected to the external genitalia. The ovaries are almond-shaped and -sized organs located in the pelvis, situated just below the division of the common iliac arteries. Ova produced by the ovaries are "captured" by the fimbriae of the uterine tubes, which open into the two top angles of the uterus. The upper two-thirds of the uterus—the body—rests on the bladder and has a very thick wall of smooth muscle. The lower third—the cervix—is a cylindrical muscular tube that protrudes into the vagina. The lumen of the cervix is a spindle-shaped canal. The vagina is a fibromuscular tube that runs from the cervix to the vestibule of the vulva.

Female Reproductive System —anterior view

1. _____

2. _____

3. _____

4. _____

Female Reproductive System— sagittal view

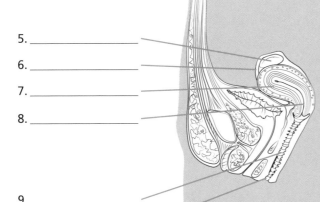

5. _____

6. _____

7. _____

8. _____

9. _____

10. _____

Answers

1. Uterine tube, 2. Ovary, 3. Ovary, 4. Uterus, 5. Vagina, 6. Ovary, 7. Uterine tube, 8. Uterus, 9. Cervix, 10. Vaginal opening to vulva

Female Reproductive System and Organs

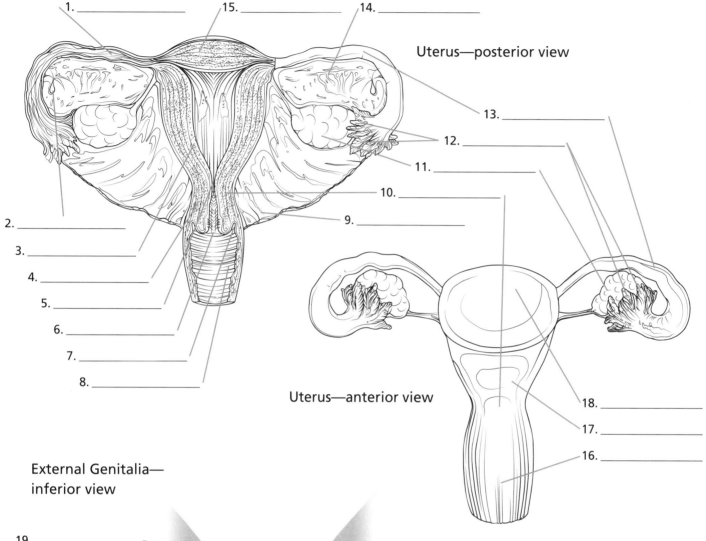

1. _____ 15. _____ 14. _____

Uterus—posterior view

13. _____

12. _____

11. _____

10. _____

9. _____

2. _____

3. _____

4. _____

5. _____

6. _____

7. _____

8. _____

Uterus—anterior view

18. _____

17. _____

16. _____

External Genitalia—
inferior view

19. _____

20. _____

21. _____

22. _____

23. _____

24. _____

25. _____

The uterus is located in the pelvis. The uterine wall consists of the endometrium, myometrium, and perimetrium. The cervix is a cylindrical muscular tube that protrudes into the vagina and is surrounded by four fornices. The vagina is a fibromuscular tube that runs from the cervix to the vestibule of the vulva. The uterus is supported by the cervical ligaments, broad ligament, and pelvic floor muscles. The female external genitalia consist of paired folds, the labia majora. These join and continue superficially to the symphysis pubis as the mons pubis (mons veneris). The labia minora are folds within the labia majora that lie on either side of the vestibule containing the vaginal and urinary openings and the openings of the greater vestibular glands (Bartholin's glands).

Answers

1. Uterine tube (cut open), 2. Hydatid of Morgagni, 3. Endometrium, 4. Myometrium, 5. Internal os of cervix, 6. External os of cervix, 7. Vaginal fornix (lateral), 8. Vagina, 9. Broad ligament, 10. Cervix, 11. Ovary, 12. Fimbriae, 13. Uterine tube, 14. Mesosalpinx (of broad ligament), 15. Fundus (of uterus), 16. Vagina, 17. Body (of uterus), 18. Fundus (of uterus), 19. Mons pubis, 20. Clitoris, 21. Orifice of urethra, 22. Labium majorum, 23. Orifice of vagina, 24. Labium minorum, 25. Hymen

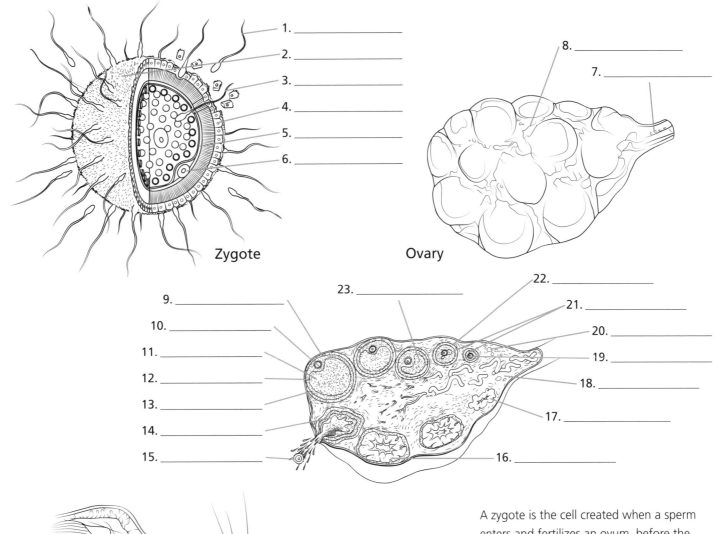

1. _____
2. _____
3. _____
4. _____
5. _____
6. _____

Zygote

8. _____
7. _____

Ovary

23. _____
22. _____
21. _____
20. _____
19. _____
18. _____
17. _____
16. _____

9. _____
10. _____
11. _____
12. _____
13. _____
14. _____
15. _____

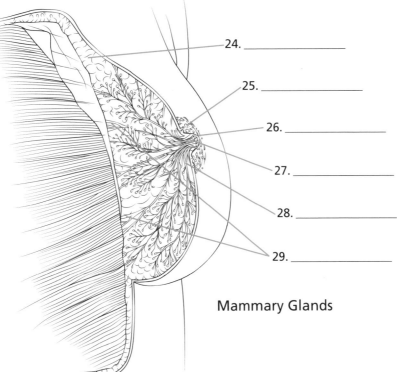

24. _____
25. _____
26. _____
27. _____
28. _____
29. _____

Mammary Glands

A zygote is the cell created when a sperm enters and fertilizes an ovum, before the process of division begins. Beneath the surface epithelium of the ovary is the cortex, which encloses the medulla at its core. The bulk of the ovary is a supporting structure called the stroma. The cortex contains ova at different stages of development. The breasts, or mammary glands, are located on the chest. They are composed mainly of adipose (fat) cells, interspersed with saclike lobules. These lobules are glands that can produce milk in females when stimulated by hormones such as prolactin. The lobules empty into a network of ducts that transport milk from the lobules to the nipple. During puberty, the sex hormone estrogen causes the breasts to grow in size, due to the laying down of fat deposits.

Answers

1. Sperm, 2. Ovum, 3. Sperm entering ovum, 4. Zona pellucida, 5. Perivitelline space, 6. Polar body, 7. Ovarian hilum, 8. Ovarian surface epithelium, 9. Mature (Graafian) follicle, 10. Mature ovum, 11. Follicular fluid, 12. Theca interna, 13. Theca externa, 14. Discharging follicle (ovulation), 15. Ovum, 16. Corpus luteum, 17. Corpus albicans, 18. Ovarian stroma, 19. Primary oocyte, 20. Primary follicle, 21. Thecal cells, 22. Primordial follicle, 23. Antrum, 24. Adipose tissue, 25. Areola, 26. Nipple, 27. Lactiferous sinus, 28. Lactiferous duct, 29. Fibrocollagenous septa (Cooper's suspensory ligament)

Fetal Skull Development

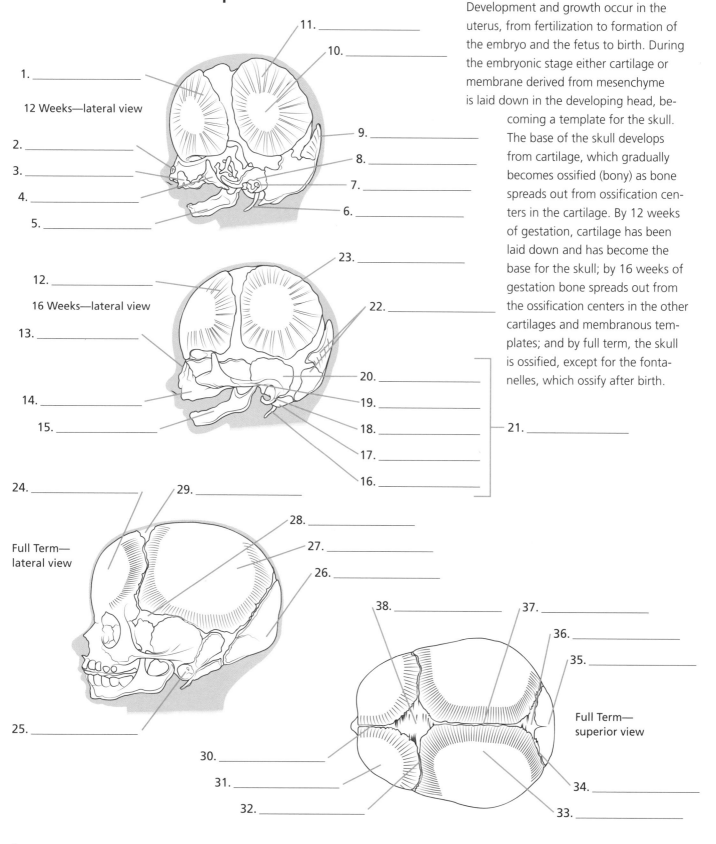

Development and growth occur in the uterus, from fertilization to formation of the embryo and the fetus to birth. During the embryonic stage either cartilage or membrane derived from mesenchyme is laid down in the developing head, becoming a template for the skull. The base of the skull develops from cartilage, which gradually becomes ossified (bony) as bone spreads out from ossification centers in the cartilage. By 12 weeks of gestation, cartilage has been laid down and has become the base for the skull; by 16 weeks of gestation bone spreads out from the ossification centers in the other cartilages and membranous templates; and by full term, the skull is ossified, except for the fontanelles, which ossify after birth.

1. _____
2. _____
3. _____
4. _____
5. _____

12 Weeks—lateral view

11. _____
10. _____
9. _____
8. _____
7. _____
6. _____

12. _____

16 Weeks—lateral view

13. _____
14. _____
15. _____

23. _____
22. _____
20. _____
19. _____
18. _____
17. _____
16. _____
21. _____

24. _____
29. _____

Full Term—
lateral view

25. _____
30. _____
31. _____
32. _____

28. _____
27. _____
26. _____

38. _____
37. _____
36. _____
35. _____

Full Term—
superior view

34. _____
33. _____

Answers

Bone Development

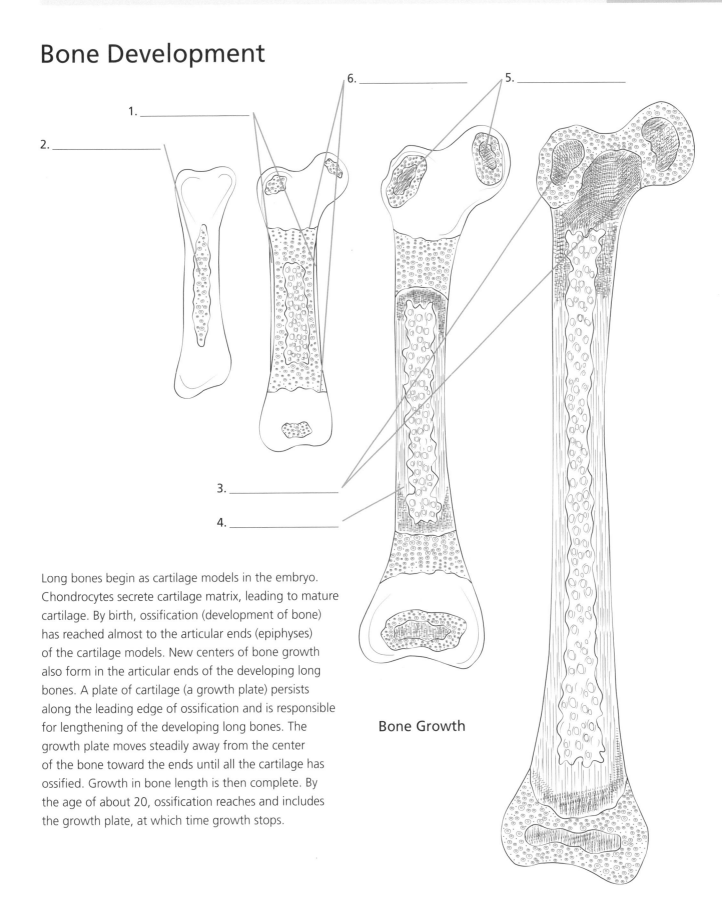

1. _____

2. _____

6. _____

5. _____

3. _____

4. _____

Bone Growth

Long bones begin as cartilage models in the embryo. Chondrocytes secrete cartilage matrix, leading to mature cartilage. By birth, ossification (development of bone) has reached almost to the articular ends (epiphyses) of the cartilage models. New centers of bone growth also form in the articular ends of the developing long bones. A plate of cartilage (a growth plate) persists along the leading edge of ossification and is responsible for lengthening of the developing long bones. The growth plate moves steadily away from the center of the bone toward the ends until all the cartilage has ossified. Growth in bone length is then complete. By the age of about 20, ossification reaches and includes the growth plate, at which time growth stops.

Answers

Index

Major topics are indicated with **bold** page numbers.

Acknowledgments

The Publisher would like to thank Barbara Krumhardt, PhD, (Lecturer in Anatomy and Physiology in the Department of Genetics, Development and Cell Biology at Iowa State University, Ames, Iowa, USA) for her expert guidance during the production of this book; David Boehm for the cover photography; Dannielle Doggett and Paula Kelly for their help in the production of this book; Harmeet Kaur and Indu L.P. from Thomson Digital for their help with the illustrations; and Lynn Lewis for her help during the conceptualization process prior to production.

The Publisher would like to acknowledge Emeritus Professor Peter Baume AO MD BS (Syd) HonDLitt (USQ) FRACP FRACGP FAFPHM, Dr. William Currie BSA MSc PhD, John Frith MBBSBSc(Med) DipEd MCH, Laurence Garey MA DPhil BM BCh, David Jackson MB BS BSc (Med), Gareth Jones BSc (Hons) MB BS DSc CBiol FIBiol, David Tracey BSc PhD, and Dzung Vu MD MB BS DipAnat CertHEd for their contribution prior to the production of this book.